50 Ways to Rewire Your Anxious Brain

Simple Skills to Soothe Anxiety & Create New Neural Pathways to Calm

Catherine M. Pittman, PhD
Maha Zayed Hoffman, PhD

New Harbinger Publications, Inc.

Publisher's Note

This publication is designed to provide accurate and authoritative information in regard to the subject matter covered. It is sold with the understanding that the publisher is not engaged in rendering psychological, financial, legal, or other professional services. If expert assistance or counseling is needed, the services of a competent professional should be sought.

New Harbinger Publications is an employee-owned company.

New Harbinger Publications, Inc.
5720 Shattuck Avenue
Oakland, CA 94609
www.newharbinger.com

Cover design by Amy Shoup; Interior design by Michele Waters-Kermes; Acquired by Jess O'Brien; Edited by Max Sylvia

Library of Congress Cataloging-in-Publication Data on file

Printed in the United States of America

25 24 23

10 9 8 7 6 5 4 3 2 1 First Printing

Catherine would like to dedicate this book to William H. Youngs, whose loss is felt in so many ways. As a mentor, supervisor, colleague, and coauthor, Bill brought so much wisdom, insight, and laughter into her life. The friendship, clinical experiences, and lunches shared with Bill made Catherine a better psychologist and professor, and he left us too soon.

Maha would like to dedicate this book to her husband Christian, her parents Yacoub and Rozette, and all the clients she has had the pleasure of working with over the years.

Contents

part four: Resisting Cortex Traps

Introduction

If you find yourself trying to cope with anxiety or panic, daily life can be a real challenge. Anxiety and panic are not only distressing and limiting, but also, others really don't understand what you are coping with. Fortunately, when coping with anxiety, you can be thankful for one thing: scientists understand more about the neurological causes of anxiety-based disorders than any other psychological diagnoses. That means we know what is happening in the brain and can offer you some answers. Unfortunately, what we know about anxiety in the brain is not always translated into tools that people coping with anxiety can apply to their daily lives. In this book, two therapists with decades of experience treating anxiety disorders will provide you with daily exercises that translate the causes of anxiety in the brain into clearly explained practices and strategies to rewire an anxious brain. The exercises provided will not only combat the distress and limitations anxiety causes in your daily life but will also make lasting changes in your brain that make it more resistant to anxiety. As therapists, we understand that lasting changes are what makes the strategies worthwhile.

The two areas of the brain most involved in the creation of anxiety are the *amygdala* and the *cortex*. Each area learns in a completely different way. The exercises in this book are designed to provide strategies that teach each area to respond in *new* ways, freeing you from anxiety. One group of exercises focuses on calming the amygdala, the part of the brain that creates the emotional and physical aspects of fear, panic, and anxiety. A second set of exercises explains strategies for rewiring the amygdala, teaching it to respond differently. A third set of exercises focuses on calming the cortex, so that it does not activate the emotional and physical responses from the amygdala. Finally, a fourth set of exercises helps you resist the cortex traps that often increase anxiety. The brain needs experiences to learn and change, and this book offers exercises that you can use in brief daily sessions. Many of the exercises suggest you use a journal to write down thoughts or ideas, so we recommend you obtain one to use regularly as you work through this book. The journal helps you tailor the strategies to your personal goals and concerns.

part one

Calming the Amygdala

1 Your Watchful Alarm System

Although the amygdala is a part of your brain small enough to fit in the palm of your hand, it has connections to various parts of the brain that allow it to have a powerful influence in our lives. The amygdala produces the defense response, a complex bodily reaction with many dimensions that is often called the "flight-or-fight" response. It also decides *when* to produce the defense response, and this can often lead to the defense response (and anxiety) being produced at times when it is not relevant or needed. Maybe you have already learned about this process, perhaps from *Rewire Your Anxious Brain* or another book, but let us briefly review how the amygdala decides to produce the defense response, and how we experience it.

The amygdala is both a watchful observer of what is happening in your environment and an alarm system that causes you to feel a sensation of danger if it detects something it considers a threat. The brain is organized to provide the amygdala with incoming information from our senses, from our eyes, ears, and sense of touch, very quickly and before the information is processed by our cortex. Because the cortex is the part of the brain that makes you aware of the information coming from your eyes, ears, etc., this means your amygdala gets information before you do. Yes, your amygdala hears, sees, and feels things a fraction of a second before you can! That is a strange thing to think about.

The brain is wired this way so the amygdala can provide defensive reactions very rapidly, but the amygdala is responding to raw sensory information and does not always have a clear idea of what it is responding to. For example, your amygdala may respond to that dark tangle of hair in the shower as if it were a spider and produce the defense response. Your heart pounds, your muscles tense, and you jump back in alarm before the visual information gets processed in your cortex a fraction of a second later, and you see the details to recognize a tangle of hair. The amygdala will also receive the hair tangle information from the cortex, and the amygdala will stop producing the defense response, but it will take some time before your body returns to a completely relaxed state. For example, your heart rate will be more rapid, and the adrenaline released will still be influencing your body for several minutes.

The amygdala can help us respond quickly. Perhaps it has saved your life on the freeway when you reacted quickly to a vehicle moving into your lane. The amygdala can be a lifesaver by producing rapid reactions in your body before you have time to think, but remember that it can also respond when no danger exists.

The Amygdala Can Be Wrong

In your journal, take a moment to reflect on your experiences with your own watchful amygdala. Can you think of a time when you were startled or frightened by something that turned out to be harmless? Maybe your amygdala reacted to a sound, a sight, or a touch of some kind, but the reaction was not needed because no danger was present. Can you think of situations in which your amygdala reacted to something as though it posed a danger, and your amygdala was mistaken? Have you noticed this happen to others? Write down some examples in your journal.

Do these situations help you recognize that the amygdala does not always interpret situations correctly? Be careful not to assume the amygdala is always accurate about the danger it responds to. The amygdala is wired to react quickly, not always accurately.

Certain emotions are often experienced when the amygdala activates the defense response: dread, fear, anxiety, or panic. These emotions are very real, but can you see how it is possible for a person to have real feelings such as dread or fear when no danger exists? This awareness—that the feeling of being in danger is not always correct—is a useful lesson to keep in mind in the process of learning new ways to respond to your amygdala.

Knowing that the amygdala does not always see or interpret things correctly is important to keep in mind. You can't always trust your amygdala. This is the first step on the path to learning how to tame your amygdala.

2 Communicating Calm to the Amygdala

Once scientists recognized the role of the amygdala in anxiety, we were able to find new ways to manage anxiety. First, it helped that we knew the location in our brain where the feelings of anxiety, fear, dread, and panic were initiated. Once we knew where it was, we could learn how it reacts. Brain imaging techniques allowed us to observe when the amygdala was activated and when its activation decreased. We could put a person in a functional MRI machine and observe how the amygdala reacted to different interventions.

Finding ways to calm the amygdala was such a relief! As therapists, we regret to inform you that nothing we tell you about your anxiety has much effect on the amygdala. The amygdala does not learn from logical explanations, like "This audience is friendly, and you have no reason to fear speaking in front of them." Similarly, when your friends and family say, "Just calm down. Everything will turn out all right," the amygdala does not stop producing the defense response. You probably know this from experience. Even what you've learned in this book so far has had no direct effect on calming or changing your amygdala.

But there are ways to calm the amygdala! Surprisingly simple and inexpensive interventions been shown to make fairly rapid changes occur in the amygdala. Perhaps the most surprising one to most people is slow, deep breathing techniques which have been shown to calm the amygdala faster than alprazolam. When a person uses specific breathing techniques while in an fMRI machine, we can actually see a reduction in activation of the amygdala. These changes, which occur within minutes after the person has begun deep breathing (Goldin and Gross 2010; Taylor et al. 2011; Zelano et al. 2016), occur faster than a medication can be digested and transported through your blood stream up to the amygdala. The problem is that few people believe that engaging in an activity as simple as breathing can be helpful with anxiety. Now that we know the importance of the amygdala in producing anxiety, and we can watch the amygdala in action, we have learned how helpful deep breathing can be.

Breathing is so simple that many people are skeptical and miss the opportunity to make use of this free and readily available resource. How can a simple process like deep breathing calm the amygdala? Breathing techniques help turn off the defense response by turning on the parasympathetic nervous system. They start processes that counter the sympathetic nervous system. The sympathetic part of our nervous system is associated with producing the fight-or-flight response, making the body ready for action, while the parasympathetic part allows the body to rest and digest. The shift from sympathetic to parasympathetic activation in the body, which you can accomplish through deep breathing, decreases the excitability of the amygdala (Jerath et al. 2015).

Deep Breathing to Calm the Amygdala

- Sit in a relaxed position. Take in a slow deep breath. It isn't necessary to breath specifically through your nose or your mouth, just whatever is comfortable. Holding your breath is not recommended. Transition directly from inhaling to exhaling when your lungs feel completely full.
- Breathe in slowly, filling your lungs completely, and if you do so, your stomach (not your chest) should expand as your diaphragm is pushed downward to allow the lungs to fill.
- Exhale slowly, perhaps pursing your lips. Emphasizing a *complete* exhale is most important in activating the parasympathetic system.
- Slow and deepen your breathing to about five breaths (including inhale and exhale) per minute.

Don't think of slow, deep breathing as "turning off" the amygdala. The amygdala is full of dynamic processes and can be activated again by various situations or thoughts. Think of breathing exercises working more like air conditioning does in your home: the air may need to come on again from time to time to cool everything down. The amygdala activates the sympathetic nervous system to support the defense response, which you are countering by activating the parasympathetic nervous system. Use deep breathing when you feel the defense response or anxiety ramping up, but also use it repeatedly during the day to reset your general stress level by activating the parasympathetic response. Sometimes it only takes three to five minutes to feel a reset process occurring, and sometimes you need to take at least fifteen minutes to reestablish relaxation. Don't underestimate the usefulness of deep, slow breathing. Remember that it gives you a way to communicate a message to the amygdala, in language it understands, by initiating a parasympathetic response that produces relaxation in the body. When the body relaxes, amygdala activation decreases.

3 Complete the Defense Response!

Jean had most of her bills set up to be automatically paid through her checking account. But one month she discovered that a payment for her life insurance policy had not been paid for some reason. She received a notice that her policy had been canceled and became quite alarmed. She had been paying into this policy for years, had invested a great deal of money in maintaining it, and knew that (at her age) purchasing a new policy would be more expensive and unlikely to give the same benefits. Jean began to imagine the difficulties her family would have without the policy in place.

She called her bank to ask why the payment had not been sent. She was transferred to several people to determine what had happened. Then she had to call the insurance company to see if there was a way to reinstate the policy. During this process, she felt a great deal of stress. She was placed on hold several times during which she worried they would not be able to help her. In the end, the error was corrected, and the policy was reinstated. Jean eventually hung up the phone and told herself that the situation was handled. The crisis was over.

But Jean's amygdala had activated the defense response and the defense response was not over! Jean's body and mood had been strongly affected by the situation. Her heart rate and muscle tension had increased, and she felt irritable. She didn't feel like eating anything and found it difficult to focus. This was because her body had been prepared to respond physically in some way, and talking on the phone didn't require that. Stress and anxiety remained in her body.

If you face a threatening situation, whether it is a spouse who is angry at you or a flat tire you discover as you plan to drive home from work, remember that your amygdala is likely to produce the defense response. Even after the situation is resolved, it can help to complete the defense response cycle by engaging in some physical activity resembling flight or fight to return your body to a relaxed state. The best way to complete the defense response is some sort of exercise, since physical exertion is what the defense response prepares your body to do. But you can also turn off the defense response by countering it with the parasympathetic activation produced by deep breathing or doing some yoga. Even getting a good night's sleep can count as resetting the body, but often it is difficult to get a good night's sleep if you have not done something more physical with your body first.

When Your Defense Response Is Activated, Do You Complete It?

During the last week or two, think of at least three times when you think your amygdala activated the defense response. For each situation, write in your journal about what happened, noting why you think your defense response was activated. What did you feel that indicated the defense response? Then note whether you did something to complete the defense response, even if it took you some time to do so. If the situation required you to complete the defense response in some way, you should count that (e.g., I walked home due to a flat tire). Note whether completing the defense response helped reduce the anxiety and stress in your body. How long do you think feelings of stress lingered?

Here are some ways to complete the defense response next time it's activated:

Aerobic exercise of some kind, including walking, dancing, climbing stairs

Tensing and relaxing muscles for several minutes

Shaking your body all over as if you are shaking water off each arm, leg, etc.

Doing some yoga

Deep slow breathing for ten to fifteen minutes

A good night of restful sleep (seven to nine hours)

Getting a massage

Getting a long hug

Snuggling with someone

Having a good laughing session

Having a good cry

Hitting a punching bag or beating a rug

Petting or cuddling a pet

Even when the situation itself may seem to be resolved, stress and anxiety remain in your brain and body unless you make a deliberate effort to complete the defense response. Finish it!

4 Relax Away the Tension

When the amygdala is activated by something, such as a growling dog, a threatening remark, or a car coming into your lane, muscle tension is one of the first reactions you may notice in your body. When the defense response is initiated by the amygdala, your body prepares to respond. Try to tune into the specifics of what happens for you personally when the defense response is activated. Do you grit your teeth, tense your stomach, or bounce your legs? These involuntary reactions make sense if you recognize that the amygdala is designed to produce fleeing or fighting, but these responses do not fit some of the situations that trigger the amygdala. When your boss criticizes your work, or you are preparing to give a presentation to a committee, the defense response preparing you to fight or flee is not particularly helpful. In addition, muscle tension can be exhausting when it is sustained and can leave you with soreness and stiffness.

Luckily, you can take steps to reduce muscle tension, and the very process of relaxing your muscles sends a message of safety to your amygdala. When you first begin using muscle relaxation strategies, try to reduce tension in all parts of your body, from your forehead down to your toes. But as you become more aware of the areas of your body that are most likely to become tense, you can focus your relaxation efforts on those specific areas. With practice, you will find that you can relax most of your body quite quickly and then give special attention to the muscle groups that need it. Relaxing your body can help calm strong emotions because our experience of emotions (which occurs in the cortex) is strongly influenced by the physical reactions we experience in our bodies. It also helps counter the amygdala-activated defense response (Shafir 2015) and reduce your anxiety (Jasuja et al. 2014).

Practicing progressive muscle relaxation is a great way to reduce tension in your body.

Progressive Muscle Relaxation

You can do this exercise lying down or sitting up. It may help to have someone read the instructions to you slowly.

Have your eyes closed or focused softly on a certain point. Start by taking five or six deep, slow breaths to help relax your body. During this entire process, continue to practice slow, deep breathing.

Now focus on tensing each part of your body for five seconds and then let go. For example, start with your forehead, your eyelids, your lips, your jaw, and your tongue. Squeeze each set of muscles for five seconds and then let go. Feel the difference between tension and relaxation.

Next focus on your neck and shoulder muscles. Tip your head back and tense your neck. To relax your neck, roll your head around your neck to the right and the left. Move to your shoulders. Tense your shoulders by raising your shoulders up to your ears and then relaxing.

Now, move to your hands and arms. Make your hands into fists and squeeze your fists. Then bring your fists to your chest to tense your arms. Then release your fists, drop your arms, and notice the difference between tension and relaxation.

Next, move to your buttocks by squeezing these muscles for a count of five and then let go. Continue to notice the difference between tension and relaxation. Make sure you are continuing deep, slow breathing throughout this exercise.

Move onto your thighs and calves. Push your feet into the floor. Squeeze your thigh and leg muscles as tightly as you can for five seconds and then let go. Notice the relaxation.

Lastly, move to your feet. Point and curl your toes under while squeezing the muscles for five seconds and then let go.

Finally, focus on letting all tension drain out of your body. Breathe deeply and slowly.

At first, practice this twice a day. After you learn which muscles get most tense, focus on tensing and relaxing only those, to complete relaxation more quickly. Then, focus on relaxing your whole body at once. Use relaxation frequently. If you are feeling tense or anxious, you can simply stop, take a few deep breaths and then tense and relax any problem muscles, getting to a state of relaxation quickly. This calms the amygdala!

5 Imagery for Relaxation

When the amygdala activates and you feel stress in your body, shifting attention away from the threat can help you create calm. We have shared various strategies for making changes in your body that promote relaxation because when the body is calm, a message of calm is also communicated to the amygdala. Using guided imagery is another strategy that can help calm the body and the amygdala. You can never have too many strategies for creating calmness, and guided imagery is a very pleasant one. Imagine being in a stressful public space but being transported elsewhere through your headphones...

Guided imagery or visualization is a technique in which you practice visualizing yourself in calming locations like a peaceful beach or sunflower field. The visualization encourages your body and mind to enter a relaxed state. Research suggests that guided imagery helps reduce anxiety and promote relaxation (Menzies et al. 2014). Even though you don't specifically focus on each muscle group, imagining yourself being in a calming setting provides relief from bodily tension and amygdala-activating thoughts.

This approach is especially useful for people who have a creative imagination. Anxious people who are creative thinkers, and who have a strong ability to imagine situations in detail, are often very gifted at worsening their anxiety by imagining scenarios that produce a great deal of distress. Are you very skilled at frightening your amygdala by imagining disastrous situations and possible catastrophes? If so, you should also be able to use your imagination to visualize more calming scenarios and allow yourself to experience escape and relief from anxiety. (If you are less skilled at visualizing situations, you may find the use of guided imagery more challenging, but then, luckily, you also probably can't terrorize yourself with your imagination either!)

Initially, it is useful to have someone read you a guided imagery scenario like the one presented below. When someone else reads the description, you can close your eyes, and concentrate on imagining yourself in the situation. You can also read and record the scenario and play it back for yourself. Finally, if guided imagery works well for you, you can find a variety of guided imagery recordings online.

Guided Imagery

Spend ten minutes during your day visualizing a peaceful setting. Sit comfortably in a chair and close your eyes. You may ask someone to read the following paragraphs to you, slowly, to allow you to focus on each sentence.

Take in a few deep breaths. Start to relax as you imagine yourself standing and gazing at a peaceful beach. Imagine the waves gently coming up to shore. Hear the sound of the water. Imagine the birds flying above and hear their occasional cries. You hear the soft sounds of the trees swaying in the wind and look over to see them. Smell the scent of the ocean water and any other smells in the air. Imagine feeling the warmth of the sun on your skin.

Imagine taking off your shoes. Walk down the beach, enjoying the way the warm sand feels on your feet and between your toes. Listen to all the sounds, the water, the birds, the children playing nearby. Step into the shallow waves lapping at the beach. Feel the cool water wash over your feet. Use all your senses to imagine this peaceful, quiet scene. Allow yourself to continue to explore the beach for a few minutes.

When you are finished, ask yourself how you felt. What did you notice? Were you able to immerse yourself in the setting? Do you feel more relaxed? Could you choose a scene to do yourself?

The setting does not matter if you would prefer something different. You can use your journal to write down different scenes you would like to focus on. What is important is that you imagine the sights, sounds, and smells of the scene in detail. This helps you engage deeply in the setting through multiple dimensions of experience. It may feel challenging at first if you are not used to visualizing. Some people share that they do not have much of an imagination. That is ok; this may not be your preferred method of relaxation. But if you are good at visualizing, make sure you use the power of imagination to relax yourself more than to activate the amygdala.

Many guided imagery scripts for relaxation are available online. We all have a favorite place to visit! Look for the specific imagery that appeals to you. Instead of letting anxiety control your thoughts and mind, you can take charge by shifting attention to a guided imagery or visualization.

6 When Your Amygdala Gets It Wrong

When the amygdala is triggered, it's a very real experience. The amygdala creates changes in the body that produce physical sensations and are interpreted by your cortex as emotional reactions. This can feel overwhelming! But it is important to remember that the amygdala very often creates these sensations even when we are not actually in danger. We simply experience the results of the amygdala trying to make sure we are protected, *should danger occur.*

The amygdala has been passed down to us from our ancestors and is prone to errors. We are the descendants of the frightened people whose amygdala activated their bodies to respond quickly to danger. In our past, humans benefitted from having an amygdala that assumed it's better to respond as if something is dangerous and be mistaken, rather than respond as if something is safe and face a threat unprepared for action. So, the amygdala is prone to seeing more situations as dangerous than is the case. This means you can't always trust the amygdala when it activates the defense response because it will frequently respond to safe situations as if they are dangerous.

The amygdala can also be wrong because it is wired to activate the sympathetic nervous system and prepare the body for fleeing, fighting, or freezing. Perhaps these behaviors were helpful in our prehistoric past, but they do not always fit in our twenty-first-century world. Running or fighting doesn't help in many stressful situations that we are confronted with. In fact, when the defense response activates, it often interferes with more suitable responses. For example, talking over a problem with someone can often help manage a situation, but calm talking is difficult when your sensations and emotional reactions are more consistent with fighting or fleeing. Even freezing can't provide much help... So, we must deal with an activated body and emotional system that does not fit the situation.

Observe Your Defense Response

Today, observe your own reaction when the amygdala produces the defense response. Whether it is a startle response, a feeling of dread, or even a bit of panic, notice how you are thinking about the sensations. Because you feel a sense of threat, are you assuming that there is a danger? Are you thinking that the sensations that you have (including your emotions) must be taken seriously? Do you have thoughts about health problems, like having a heart attack? Do you take your experience to indicate that something bad is likely to happen? Do you begin to imagine distressing events occurring in your life? Write your responses in your journal.

Also, write about if you feel as if you must do something. Do the changes you experience in your body lead you to take actions that feel necessary but are not in your best interest? Remember that fleeing, fighting, or freezing are the limited responses of the amygdala, but are not the only options that you have. Take time to simply observe, with curiosity and interest, the impulses you are experiencing, resisting the tendency to allow the amygdala to take charge of your reactions. (Note that, if you are having a full-blown panic attack, this kind of mindful reflection is not typically possible.)

Your thoughts about the way your body responds to anxiety can lead to even greater distress. Be aware of whether you are *assuming* a danger exists when it may not. Remember the amygdala is often wrong when creating the defense response. Sometimes no threat of any kind exists. Sometimes the situation you are worrying about never occurs. In other cases, some kind of threating situation does exist, but the reactions that the amygdala is capable of creating (fight, flight, or freeze) are not appropriate or necessary in the specific situation. Keep in mind that you can simply observe how the amygdala influences how your body and mind react and think to yourself, *Thanks for trying to protect me, amygdala, but I'm not getting caught up in your reaction*. You can use twenty-first century strategies in response to the situation, including deep breathing, exercise, a change in focus, a discussion, or planning. These give you much more flexibility in responding than the amygdala is capable of.

7 Getting the Sleep the Amygdala Needs

When a person gets little sleep, the amygdala is more likely to be reactive. Maybe you've noticed that when you get little sleep you are more anxious and irritable. But now that we have brain scanning techniques, we have been able to document a connection between the amygdala's activation and the amount of sleep a person gets. Even one night of sleep deprivation can make the amygdala react more strongly (Yoo et al. 2007). If you make an effort to get the sleep that the amygdala needs, you will find you are much calmer.

Research has shown that not only does the *amount* of sleep we get influence the amygdala, but the type of sleep does as well. When we sleep, we go through a series of different stages. In each stage, we experience a different type of sleep, and we go through the stages according to a common pattern. The most important stage of sleep for the amygdala, based on the research, is a stage of sleep called Rapid Eye Movement (REM) sleep. The less REM sleep a person gets, the more a person experiences activation in the amygdala (Prather et al. 2013). During the REM stage of sleep, we see the amygdala being activated, so this stage clearly has a connection to amygdala functioning (Sleep Foundation 2022).

REM sleep occurs very little in the first hours of sleep, often not beginning until after 60 to 90 minutes of sleep (Sleep Foundation 2022). The longest periods of REM sleep occur in the late hours of sleep, making elongated periods of sleep necessary for us to experience enough REM sleep for the amygdala's needs. For example, in the first four hours of sleep, you get very little REM sleep compared to what you get in your sixth and seventh hour. Having a sixth and seventh hour of sleep is essential for people who want to calm the amygdala.

How to Prioritize Sleep

1. *Make a commitment to sufficient sleep, as if it is an important part of reducing anxiety, because it certainly is!*

2. *As much as possible, have a regular bedtime, even on the weekends. The brain gets accustomed to a certain schedule and changes can disrupt the sleep cycle.*

3. *Don't use alcohol for help in falling asleep. It delays how quickly you get into REM sleep and reduces the amount of time you spend in REM sleep periods.*

4. *Avoid caffeine for six hours before your bedtime. Caffeine also interferes with normal sleep processes, even if it does not prevent you from falling asleep.*

5. *Engage in a regular exercise routine, at least walking, for at least three or four days a week. This improves your ability to get extended sleep.*

6. *Sleep in a cool, quiet, and dark location, and minimize interruptions. The impact of pets and*

family members who interfere with your sleep should be considered.

When we encourage you to prioritize sleep, we are fighting against a great deal of cultural pressure. Particularly in the United States, sleep is often treated as if it is a waste of valuable time. Perhaps this attitude comes from an inherited Protestant work ethic, but regardless, a common impression is that sleep wastes time and can be limited. Most people assume the brain is inactive during sleep; on the contrary, important brain activities occur in the brain during sleep, such as the production of chemicals needed in daily brain functioning. Sleep is important for a variety of essential functions, including immunity, memory, and mood.

Getting a long period of sleep each night (seven to eight hours) will help you have a calmer amygdala. Tracking your hours of sleep in your journal might help you see a pattern that increases anxiety. If you have a job or sleeping situation that interferes with your ability to get sufficient sleep, you may need to make a change. Sleep is an inexpensive form of treatment for anxiety that benefits you in many ways. Evidence shows that brain processes that occur during REM sleep are associated with a reduction in amygdala activation the next day (van der Helm et al. 2011). Consistent with this, sleep deprivation has been proposed to contribute to panic attacks (Babson 2009), and we have seen clients who recognize that when they have a panic attack, it is often after a period of insufficient sleep. If you start each day with sufficient REM sleep, you have a head start on having a calm day.

8 Laughter Is Good Medicine

People who struggle with anxiety tend to take life very seriously. It's hard to relax and let your guard down. Also, your thoughts may be focused on concerns and dominated by worries. You may be so preoccupied with anxious thoughts that it's hard to stay in the present moment. This is common when your brain is being strongly influenced by the amygdala. When the amygdala is activated, your guard is up, your thoughts are focused on potential threats of various kinds, and your body is tense. It's not surprising, then, that you aren't likely to find things funny!

But if we allow the amygdala to dominate our thinking processes, we will miss opportunities to enjoy our lives. Remember that the amygdala's approach to concerns is to create the defense response, to prepare you to fight, flee, or freeze. While these responses were valuable to our ancient ancestors, they are not as useful in the twenty-first century. Recognize that you are not consciously responding this way; the amygdala is trying to take control to protect you. But you can take steps to shift your perspective. Don't fall into the trap of behaving as if the amygdala is right. Instead, try to lighten up and find humor in your life.

A good laugh changes your breathing and makes you breathe deeper, which relaxes you. Joking and being playful also shifts your attention away from a focus on threats and increases your connection with other people. Humor has a way of decreasing nervousness between people and promoting a feeling of playfulness that can lighten the mood. Evidence even shows that laughter attracts the amygdala's attention; the amygdala is both activated by laughter (Sander, Brechmann, and Scheich 2003) and involved in producing laughter in response to others' laughter (Lombardi et al. 2022). Yes, laughter is contagious, and the amygdala is involved.

Laughter therapy is real! "Laughter is the best medicine" because real changes occur in your body when you allow yourself to laugh—even when your body seems to be prepared for a threat. Cortisol and adrenaline are released when the amygdala produces the defense response, but when you laugh, different hormones are released, including endorphins, dopamine, and oxytocin (Yim 2016). *Endorphins* are chemicals that not only increase feelings of wellbeing, but they also reduce our perception of pain. They give you a natural boost when they are produced during certain activities like exercise, sex, eating chocolate, and—perhaps surprisingly—simply laughing. They are like natural medicine, activating the same parts of the brain that are activated when you take cocaine! What is most helpful to people suffering from anxiety is that endorphins are an antidote to the defense response that the amygdala creates (Pilozzi, Carro, and Huang 2021). Laughter decreases heart rate and blood pressure and reduces the tension in your muscles.

Dopamine is another chemical released when you are laughing (Yim 2016). This chemical is involved in our experiences of pleasure and reward and gives us a good feeling. Finally, *oxytocin*, often called the cuddling hormone, is released when you laugh, and it plays a role in social bonding (Woodbury-Fariña and Schwabe 2015). The amygdala has receptors for both dopamine and oxytocin, so it can be directly affected when they are released due to laughter.

Make Time to Laugh

What kind of humor gets you laughing? Consider your experiences and observe yourself to see what situations tickle your funny bone. Also, do you have ways that you get others laughing, and does that improve your mood? We all have our own kinds of humor, from puns to physical comedy, from cleverly told jokes, to silliness with others that is just plain goofing around. Anxious people often miss out on humor, so make sure you take advantage of opportunities to laugh and be playful. Which friends, family, and coworkers are most able to joke and get you laughing? Recognize them as helpful resources in your life. Having more humor in your daily life can help you combat stress and anxiety.

Get your journal and make a list of opportunities that get you smiling and laughing. List anything from cat videos to practical jokes, movies to silly board games. Think of people, pets, shows, and activities that get you laughing. Then make a goal to increase your exposure to these situations and put a star next to ones that you can target to make more time for. Make having an opportunity for laughing a goal for each day.

You can take steps to keep yourself from letting anxiety trap you in seriousness. Lighten up! If you find this difficult, try to find someone in your life whose humor you appreciate. Make time to be with those who are laughing or tickling your funny bone. Remember that the amygdala finds laughter to be contagious!

9 Exercise Is Just What the Amygdala Wants!

Knowing that the amygdala activates the fight, flight, or freeze response helps us make sense of our anxiety reactions. The amygdala activates a built-in defense response to protect us from threats we experience, even though running, fighting, or freezing are not helpful in dealing with most twenty-first-century problems. When you feel your heart pounding, feel dizziness as blood flow is redirected to your muscles, or experience activation in your muscles that makes it hard to stay still, you know that the amygdala has activated the defense response. Anxiety and fear reactions are very physical experiences because dozens of changes are occurring in your body to prepare you to be successful at fighting or fleeing. Similarly, when you feel strongly compelled to avoid something or escape from a situation or feel a strong urge to strike out at someone or something, the amygdala may be taking charge of your body to protect you from a threat.

When you understand what is happening in your body, you can see that some form of exercise, particularly aerobic exercise, is just what your body is prepared for. So why not take advantage of that state of activation to engage in some kind of physical activity? At first, it may not make sense to jump on a stationary bike for a few minutes after getting a stress-provoking phone call. But when you exercise you are doing exactly what the amygdala wants you to do. When you engage in some kind of physical activity, even something as simple as taking a walk, you are using the effects of the adrenaline that has been released, effects that can remain in the body longer if you don't engage in some physical activity.

We know that we can't reason with the amygdala. It does not operate based on logic. Trying to remind yourself that the presentation you need to make at the office is not a threat that requires the defense response will have little effect on the amygdala. Instead of looking for a way to convince the amygdala to calm down, you can take the kind of action that your body is prepared for. The truth is that exercise is more effective than trying to reason with the amygdala. If your amygdala thinks you should run away, then take a brisk walk. We like to joke "Just do something that can show the amygdala you have taken action, and it will calm down."

Take Action!

The next time the defense response is activated by your amygdala, engage in some kind of exercise that works for your body and ability. If you feel anxious, doing something aerobic that activates your large muscle groups is most effective at reducing amygdala activation (DeBoer et al. 2012). Some kind of moderate exercise, approved by your doctor and enjoyable to you, is the best. You can turn on music and dance for twenty minutes if you wish! You could go for a jog, take a brisk walk, go for a swim, or climb some stairs in your office building. Whatever you can do that gets you moving and promotes a moderate increase in your heart rate for at least twenty minutes can help. After vigorous exercise, you will experience reduced muscle tension, increased relaxation, and even stimulate the production of endorphins (Bourne, Brownstein, and Garano 2004). Most studies of exercise have shown that it reduces anxiety (Ensari et al. 2015; Rebar et al. 2015), often within twenty minutes (Anderson and Shivakumar 2013; Chen et al. 2019), which is faster than most medications can work. But you are not just getting relief from anxiety; you are also sending a message to your amygdala: "I've taken action! You can relax now." Then go on with your day and focus on things other than whatever triggered the amygdala.

You should know that, in addition to reducing anxiety, exercise produces direct effects in the amygdala. Exercise has been shown to produce changes in both brain chemistry and neurons in the amygdala (Christianson and Greenwood 2014), putting the brakes on its activation. Exercise also alters the amygdala's interactions with other areas in the brain (Chen et al. 2019), suggesting that vigorous exercise changes the way the amygdala sends signals to regions of the brain that are involved in producing anxiety. It would make sense that when a person performs some physical exertion, some type of mechanism in the brain would reset the amygdala, in the same way, that chemicals released in the gut and liver make the feeling of hunger go away after someone has eaten.

10 Mindfulness Meditation and the Amygdala

The amygdala is designed to hijack our attention to focus on anything it considers a threat, but meditation teaches you to control what you are thinking about and maintain your focus. Before starting meditation, you should know that this "control" of your mind is not what you might expect. The mind is not like a television, staying on a certain channel until someone changes it. The mind constantly experiences different kinds of awareness, processing sounds and sights in the environment, and experiencing memories, intentions, and other thoughts. The mind is not intended to be rigidly focused on anything, and awareness shifts frequently in any healthy mind.

Many people think that meditation is emptying the mind, but it is not the nature of the mind to be empty. You always have something on your mind. But when you practice meditation, you learn how to have more control over your mind. It's not a grasping kind of control like a hand firmly holding something. Focusing your mind is more like controlling what you look at. When you look at something with your eyes, you're not tightly gripping it and you might glance away occasionally, even when you try not to. Similarly, the mind naturally wanders, but it can be focused, by simply bringing your attention back to what you intend to focus on. Don't be discouraged if your mind doesn't "hold onto" the focus and you often need to bring your attention back to the focus you choose. This is normal.

Sounds and activities occur in the environment and thoughts pop into your mind causing you to "glance away" from your focus, but every time you bring your focus back, you are strengthening your ability to be in control of your focus. Just like building up muscles with repetitive exercise, you can strengthen your ability to focus. Remind yourself that, like your eyes, it is not the nature of the mind to grasp firmly. If it wanders ten times, bring it back ten times.

Mindfulness meditation is one of the most fundamental types of meditation. Being mindful means focusing on awareness of your experience, starting with your senses. Start by focusing on one sense at a time. For example, take a few minutes to listen to every single sound that you can hear in your current location. You suddenly become aware of all kinds of sounds that you were not focusing on before. The heating system, the clock ticking, your own breath... You can learn from mindfulness meditation that you can take control of your ability to focus relatively easily when you focus on your senses.

Focusing on your senses also brings you into the present; what you see, hear, smell, feel, or taste connects you with the present moment, something that can help us get away from worrying about the future, or dwelling on past difficulties. Focusing on worries or past concerns will activate your amygdala to produce anxiety, increasing your blood pressure and feelings of distress, when you are actually in a safe place. Instead of focusing on thoughts and images that cause the amygdala to produce a defense response, shift to listening to the air circulation and the amygdala will calm down.

Mindful Meditation Calms the Amygdala

We often begin mindfulness practice by focusing on some food. A piece of candy or fruit can be used, but any food will do. Get yourself some food of your choice. Let's imagine that you are using a peppermint candy. We will ask you to try to use all your senses to practice mindfully experiencing this candy. Starting with your vision, take a close look at the candy. Is it wrapped in something? Notice any colors or writing. Examine the wrapping carefully. One aspect of mindfulness is to be open to new experiences and details that you might not have attended to before.

If you need to unwrap the candy, notice not only what you see, but the sounds that you hear. Then take time to look at the candy itself, to notice details. Shift to experiencing the smell. Then put it in your mouth to taste it. Notice the taste, including how different parts of your tongue taste it. Explore how it feels on your tongue. Eventually bite it and notice how the feel and taste changes when it is broken up.

It's amazing how mindfulness can refocus the mind. When you feel like you can't help focusing on what makes you anxious, try mindfulness. Mindfulness meditation redirects your thinking in a given moment, and, each time you practice it, you are developing and strengthening the circuitry that helps you shift your focus. You strengthen your ability to direct your thinking to what *you* want to focus on.

11 The Shape of Anxiety

Anxiety is what we feel when the body is in defense response mode. The amygdala is activated and produces unpleasant bodily responses intended to be protective. Fight, flight, or freeze responses show up in the body as anything from muscle tension to a pounding heart. The emotions range from a sense of dread to complete panic. But it can help to look at the "shape" of anxiety. The shape of anxiety is how it would look on a graph showing when anxiety increases and decreases over time. The shape that anxiety takes, the pattern that it follows, may surprise you. Anxiety doesn't always peak when you would expect!

Most people believe that anxiety is going to follow a logical pattern. They have an idea of how anxiety will increase when they face a triggering situation, whether it be having to give a toast at a wedding, flying on an airplane, or dealing with a pile of bills they can't pay. They anticipate that anxiety will increase steadily as they approach the situation and rise to an intolerable level as they enter the situation. They expect that, in the situation, anxiety will be maintained for an unbearable period of time, unless they leave the situation. This sounds very discouraging, but fortunately, it is not correct.

The shape of anxiety is such that it peaks *just before* we face the trigger. It makes sense to assume that our anxiety will be the worst when we actually encounter the trigger, not before. But when you understand that anxiety is part of the amygdala preparing for flight, fight, or freeze, you realize why that would not work. The amygdala operates on the need to react *before* we encounter a threatening situation. Think of it this way: if the amygdala waits until you are in the tiger's mouth to react, that will be too late. The amygdala needs to produce an anticipatory reaction, designed to help you escape, fight, or remain motionless to prevent detection. As a result, the pattern anxiety follows is to be strongest before you encounter the trigger.

Consider the situation of giving a speech before a group of people. The shape of anxiety is that it rises as the time for the speech approaches, peaks as you are getting ready to step before the audience, and—surprisingly—seems to decrease after you begin speaking. This is the anticipatory nature of the fight, flight, or freeze response. People assume that their anxiety will be maintained once they are in the presence of a trigger, but if no danger appears, the amygdala typically calms down. There are times when this is not the case. In part 3, we will discuss how the cortex can cause the amygdala to maintain activation by producing threatening thoughts. But if the cortex focuses on calm thoughts, the amygdala will recognize that danger did not appear, and the fight, flight, or freeze reaction will decrease as you remain in the presence of the trigger.

Understanding the shape of anxiety can help you avoid the mistake of thinking that the anxiety you feel *before* facing a trigger will be even worse once you actually face it. A client facing his fear of flying said, "If I feel this scared before I board the plane, how bad am I going to feel during the flight?" and was stunned to learn he will actually feel *less* scared. Think of it this way: the amygdala is trying to convince you to run away. But if you don't run away and you face the situation, the amygdala will recognize the absence of danger and calm down. Some physical sensations, like muscle tension, may ease right away while others may take a bit of time, but as the effects of the released adrenaline wear off, you'll generally feel much calmer than you would expect. Your peak anxiety is typically *before* you enter the situation, not when you are in it.

Consider the Shape of Your Own Anxiety

Examine your own experiences of anxiety. Can you see your anxiety peaking before the situation, not in the situation? Consider both minor anxiety-producing situations (having others visit your home) and major anxiety-provoking situations (returning to driving after an accident). In your journal, recall the situations and rate your anxiety at different points to see if anxiety tends to peak just before you are in the situation you fear.

Anxiety is by design anticipatory. It is rooted in the defense response and can be reduced when no real threat is encountered. This knowledge can greatly assist you in pushing through anxiety.

12 Play!

When you're anxious—the amygdala is busy creating the defense response—your body is tense, you feel distressed, and your thoughts focus on dealing with some kind of threat. Spending a lot of time in this state can have a limiting effect on your life. You may have difficulty relaxing and finding enjoyment in daily activities. You may take things too seriously and allow worries to be your focus. It may seem wrong to have fun or make jokes. Being playful may seem inappropriate when you feel like a potential threat is looming, but playfulness is one of the best ways to cope with an activated amygdala.

Remember that the amygdala can easily misinterpret a situation or overestimate danger, so it should not always be trusted. If you wait until your amygdala is completely calm before allowing yourself to let your guard down and have some fun, you are giving it too much control over your life, and you'll miss out on a lot of good times! Many people don't realize that they are holding themselves back by not allowing themselves to engage in hobbies, sports, or games because somehow, they do not feel safe enough to let themselves relax. Some people believe that play is for children while adults should be serious and not spend time in leisurely pursuits. Adulthood is actually enhanced by play.

Instead of waiting until anxiety goes away, worrying about whether your friend is angry at you, dreading an upcoming meeting at work, or checking your email repeatedly, make time in your day for play! Allow yourself to play a game on your phone, ask your partner to play cards or a board game, or throw a ball for your dog. Don't wait for a feeling of complete safety or until you have reassurance that a situation will turn out the way you hope it will. Every day should include some fun.

Give yourself permission to take a situation lightly. Use humor and playfulness to cheer yourself. Get a friend or family member to laugh or clown around with you. Deliberately hang around with someone who is good at having fun. Find humor in yourself and in life. Allowing yourself opportunities to be playful, instead of living in fear, helps you create feelings of amusement and joy for yourself and others.

How to Play

Make a conscious effort to include enjoyable activities and hobbies in your daily life. Make a commitment to spending at least half an hour each day in play, but don't limit yourself to that. Start by considering what you enjoy. Do you like crossword puzzles or cross-stitch, soccer or softball, gin rummy, or Words with Friends? Are there things you once enjoyed but haven't done recently? Consider whether anything you used to do as a child still appeals to you.

Here is a list of enjoyable activities and hobbies to consider. Choose activities that entertain and amuse you. You don't have to be particularly good at them, as long as you enjoy them. Make sure that some of the activities you choose are ones you can do on your own, and don't depend on someone else being around. You can also use Google to explore "fun activities for adults." Write your favorite ideas for play in your journal.

Card games (including Solitaire)

Sports

Doodling or sketching

Puzzles

Video games

Baking

Painting

Nature walking

Board games

Gardening

Writing (poetry, etc.)

Dancing

Leisure reading

Zumba

Water activities

Storytelling

Playing with a child

Collecting

Playing with a pet

Shopping

Bird-watching

Singing

Playing an instrument

Antiquing

Listening to music

Theatre

Watching sports

Redecorating

So many times, people coping with anxiety don't allow themselves to let their guard down, remaining constantly watchful and cautious about keeping their attention on what could potentially go wrong, and what symptoms they are experiencing. All this focus on potential danger increases amygdala activation, and allows the amygdala to dominate their lives, when they could be living a much more fulfilling life with laughter and fun. Make sure that your daily life includes some play. Erik Erikson, the American psychoanalyst, said that "the playing child advances forward to new stages of mastery," but also recognized "the playing adult steps sideward into another reality." Play is a wonderful reality we all deserve to spend time in.

13 Maintaining an Amygdala-Friendly Schedule

The more you live a life that considers the needs of the amygdala, the less anxiety you will have. Using modern brain imaging techniques, we can observe how the amygdala responds in a variety of situations and which factors impact the amygdala's reactivity. We've learned that when your amygdala is reactive, you have more physical activation and anxiety, and when your amygdala is less reactive, your body and mind are calmer. So, if you practice daily routines that produce a less reactive amygdala, you set the stage for less anxiety overall.

Consider the finding that even one night of sleep deprivation can make the amygdala react more strongly (Yoo et al. 2007). Imagine what your amygdala is like when you are frequently sleep deprived! It can be so reactive! Even though your mother didn't know about the amygdala, she probably warned you that poor sleep would make you cranky. For the amygdala, good sleep is extended sleep—at least seven or eight hours—in order to achieve the periods of REM sleep that only happen after six or so hours of sleep. If you deliberately work on getting enough sleep each night, you will be amazed at how much less anxiety you have in general.

Your diet can also affect the amygdala. The amygdala has receptors that allow it to monitor blood sugar levels and is more likely to react during periods of low blood sugar (McNay 2015). Try to eat meals at regular times, and don't allow yourself to get too hungry. Protein lasts longer in your system than carbohydrates, which provide a quick burst of energy that quickly fades and then low blood sugar hits.

Speaking of diet, caffeine increases anxiety for some people. As the most popular drug in the world (Rogers 2007), caffeine is often invisible in a person's life and not recognized as impacting anxiety. But caffeine activates the sympathetic nervous system and initiates adrenaline release, so you may find it contributes to increases in heart rate and blood pressure that worsen your experience of anxiety. Also, don't forget that caffeine interferes with the feeling of tiredness and the regular sleep cycle, resulting in less sleep time and poorer sleep quality (Roehrs and Roth 2008; Watson et al. 2016). To be confident in getting good sleep, you should not ingest caffeine for at least six hours before going to bed.

Another important habit that can decrease amygdala activation is exercise. When you experience anxiety, you can reduce it by performing just twenty minutes or less of aerobic exercise (Anderson and Shivakumar 2013; Chen et al. 2019). Decreases in amygdala reactivity have been shown after only fifteen minutes of exercise (Schmitt et al. 2020).

Including exercise into your schedule for at least thirty minutes, three to five times per week, can result in lasting changes in your body and your amygdala. Regular exercise decreases a person's general anxiety level and results in less sympathetic nervous system activation (Anderson and Shivakumar 2013). People who exercised for just twenty-five minutes every two or three days, after only twelve of these exercise sessions, experienced less anxiety in general than people who were not exercising (Lattari et al. 2018). Further, regular exercise results in changes in the amygdala itself. Neurons in the amygdala become less active after regular exercise (Heisler at al. 2007). Exercises that involve large muscle groups are most helpful in reducing amygdala activation. Benefitting from exercise does not require any special training. Regular, brisk walking can result in changes in your anxiety and your amygdala.

Schedule Checkup

Answer "Agree" or "Disagree" to the following:

1. *I have a regular bedtime.*
2. *I get at least seven or eight hours of sleep each night.*
3. *I avoid drinking caffeine six hours before bedtime.*
4. *I eat regular meals, including breakfast.*
5. *I often eat sweet treats and carbohydrates for snacks.*
6. *Caffeine causes me to feel anxious or too activated.*
7. *I go long periods in a day without eating.*

8. *I often feel dizzy, shaky, or irritable due to not eating.*

9. *I engage in regular aerobic exercise at least three times per week.*

Concerns about your daily habits are raised if you answer "Agree" to 5, 6, 7, and 8. Answering "Disagree" to the remaining questions (1–4; 9) could raise concerns. If you have daily routines, consider making them more amygdala friendly.

Take time to look at your general lifestyle and consider whether it contributes to amygdala activation. When you modify your schedule to incorporate ways to calm the amygdala, life is not so stressful!

14 Responding to the Freeze Response

Sometimes the way that we respond to a stressful situation is to freeze. Freezing is part of the amygdala's fight-or-flight response, which is why we sometimes call it the fight, flight, or freeze response (LeDoux 2015). We often see the "freeze" response in other animals. "A deer in the headlights" isn't just an expression! We don't choose to be immobilized; it is a remnant of our past as prey for predators. When an animal can't escape a predator, remaining motionless to not be detected can be lifesaving.

Like other animals, we can instinctively freeze when we don't want to call attention to ourselves. In some stressful situations, we might wish we could be invisible! When the fight response is activated, but we don't feel it's safe to fight, we may freeze. You can see the benefits of freezing if you have the impulse to strike your supervisor when he speaks disrespectfully to you: Freezing won't get you fired.

When we freeze, just like when we are in fight-or-flight mode, thinking processes in the cortex are not very available to us. We can't think clearly or process what people are saying very well while in a "freeze" state. People who freeze often worry that they are "losing their mind." If you've thought this, know that disrupted thinking is a normal part of a freeze response. The amygdala is wired to make us respond this way.

Once it happens, you don't have many options, and you won't be able to do too much for a few minutes.

When you freeze, your muscles are tense, not limp. Sometimes people worry they will become unconscious, but that is a completely different response, something we sometimes see when an animal "plays dead." When you freeze, you are not going to collapse (LeDoux 2015).

Freezing often occurs when people are overwhelmed. One of the best ways to get back to feeling more like yourself is to do something simple like look around at your surroundings or deepen your breathing to shift yourself back to a more relaxed state. If you practice the steps below during a time when you are *not* in the freeze response, you'll be ready to use them if you need to. Just remember that the freeze response is intended to be protective, and that the amygdala produces it in an effort to keep you safe.

How to Manage the Freeze Response

Take some time to practice the following steps. The more you practice these, the more familiar they will feel, and you can use them to help yourself come out of the freeze response.

1. *Take some slow deep breaths and make sure you breathe out as completely as you can. The more familiar you are with slow deep breathing, the more likely you can switch to this breathing fairly automatically. This type of breathing helps*

activate the parasympathetic response that allows you to come out of freezing.

2. *Focus on looking at your surroundings; even when freezing, you are always able to move your eyes to look around. First just remind yourself of where you are. You don't have to say anything out loud but think it to yourself. Then look around and identify what you see by naming objects in the area. This can often help you move your head a bit and allow you to take stock of your surroundings.*

3. *Now that you have become more in touch with what you see in your surroundings, focus on your other senses too. What do you hear? What do you smell? Can you see or identify these things?*

4. *If you feel safe enough to move around, move your head and then your hands, and arms. This is a good sign that soon you will be able to get back to moving around normally.*

If you practice this simple routine, which is really a mindfulness practice, it can help you feel more prepared for dealing with the freeze response when it occurs.

15 Tucking In Your Amygdala

Do you often get into bed tired but have difficulty falling asleep? How can it be that you *want* to sleep but *can't* fall asleep?

Unfortunately, the amygdala is part of the problem. If we get into bed and worry about tomorrow or ruminate over what happened today, we allow our focus on potential threats to capture the amygdala's attention. As our protector, the amygdala is watchful for any dangers and if our thoughts are focused on some threat, the amygdala activates the sympathetic nervous system which can make it difficult to relax and fall asleep. Remember that we are the descendants of the worried people—the people who did *not* fall asleep at night when they recalled recently seeing a tiger prowling around the area. Remaining alert to dangers rather than sleeping saved our ancestors and left us with an amygdala that makes it difficult to sleep when our thoughts are focused on threats.

In modern life, staying awake thinking of potential threats is not likely to be beneficial. The threats we face are not prowling tigers but more likely paying our bills, conflicts with loved ones, or the demands of our jobs. Losing precious hours of sleep is actually detrimental in many ways and leaves us with a more activated amygdala the next day. So how do we put the amygdala to bed, rather than activate it?

Often, we hear clients saying that they have no difficulty falling asleep in front of the television, but when they get into bed, they just can't seem to relax. This provides a clue! When you focus your attention on information that is non-threatening, like a television show, the amygdala is less likely to keep you alert and awake.

But making the television a part of your sleep routine is not the answer. The type of blue light emitted by television, computers, or phone screens produces alerting effects and prolongs the time it takes to fall asleep by suppressing the release of the hormone melatonin, which is needed for falling asleep (Chang et al. 2015). So looking at screens at bedtime delays sleep and costs valuable REM time. Even delaying sleep for a half hour to an hour may reduce REM sleep a noticeable amount. The answer is to find a way to focus our attention on calm thoughts in other ways.

Focus on Calming (or Boring!) Information

If you listen to (not look at) your television, smartphone, or computer, and choose a program, book chapter, or podcast to listen to, you will give your amygdala something to focus on other than your worries or potentially activating thoughts. We all know how difficult it is to focus on something when someone is talking to you. So listen to something that is calming or even boring by choosing an audiobook, podcast, or program that distracts you from any amygdala-activating thoughts. Of course, you need to keep your attention focused on what you hear and not allow your focus to return to your worries. Deep, slow breathing will also help. If you find yourself still focused on activating thoughts, choose something that is more effective at keeping your thoughts off your worries. Music is not likely to be effective, for example, because we can often focus on our worries at the same time music is playing. Unless we are singing along with a song, music tends to allow us to continue to focus on our worries.

If you feel that worrying or focusing on a threatening situation is important, you need to schedule that kind of thinking for a different time of day. Give yourself thirty minutes or so to allow yourself to worry or focus on your concerns, but not at or before bedtime. You will find that you fall asleep much more quickly if you replace amygdala-activating thoughts with calming information.

Tucking in your amygdala can also be assisted by having a specific, predictable routine before bed that signals to your brain that sleep is coming. The routine can be as simple as washing your face, brushing your teeth, and putting on pajamas, or it can be more involved like taking some time to read a calming book, or drinking a cup of herbal tea. Having a comfortable bed in a cool environment and minimizing interruptions from pets or partners also helps promote lengthy, uninterrupted periods of sleep that will provide you with a calm amygdala the next day.

part two

Rewiring the Amygdala

16 Follow Your Goals, Not Your Fear

When we encourage you to push through your fears, we recognize that this is a difficult challenge. We would never suggest pushing through anxiety is easy, but the good news is that you don't need to face *all* your fears. If you are afraid of snakes, but rarely encounter snakes, that is not a problematic fear. But if you have anxiety about asking your boss for a raise, and other employees are successfully getting higher salaries while you settle for your original compensation, that is a fear that is reducing your quality of life.

When you decide to challenge yourself to push through your fears, focus on situations in which your life is being restricted or limited in some way. When the amygdala is triggered, and your anxiety prevents you from doing something that is important to you, that's when you need to put effort into teaching the amygdala to respond differently. When you stop avoiding and start facing the situations that keep you from accomplishing what you want to do, you are taking charge of your life. We want *you* to be in charge of your daily activities—not your amygdala.

So many aspects of life can be impacted by your anxiety. You may find that daily activities as simple as taking a walk or making a phone call are blocked. Your family and friend relationships may be impacted. You may not feel that you can socialize or travel with the people you care about. You may not be able to communicate well or manage conflict in your relationships. On the other hand, your relationships may be fine, but your work and your career goals are not being achieved due to anxiety or avoidance. But you don't have to be trapped by anxiety; you can make changes in your amygdala that will help you take charge of your life. We have ways to get anxiety out of your way and help you reach your goals.

Choosing Your Goals

Before facing your fears, carefully determine where you want to focus your attention. You don't need to take on all the anxiety, fear, or avoidance in your life. Instead, decide what is most important to you. You decide, not your family or friends, or even your boss. If your boss decides that you need to get over your fear of flying and begin taking business trips, you might not agree. You may not want a job that involves travel. In the same way, if your partner wants you to begin taking ballroom dancing classes, it doesn't mean the activity appeals to you. You need to carefully consider your own goals and where they are blocked. Finishing the following statements might help. Write your responses in your journal or on a piece of paper meant only for your own eyes. Note that you can write more than one response to each statement.

Your Daily Life

If it weren't for my anxiety or stress, I would want to go…

When I feel anxious, I stop trying to get myself to do activities like…

Before I had problems with anxiety and stress, I engaged in enjoyable activities like…

Your Relationships

I find myself anxious around people when they…

It stresses me out when people expect me to…

Anxiety is interfering in my relationships when I don't…

When my family (or friends) invite me, I wish I could…

With regard to a romantic relationship, I would like to…

Your Work Life

If it weren't for my anxiety, at work I could…

It would improve my performance at work if I could…

My progress at work is limited when I turn down opportunities to…

My anxiety interferes with time management at work because I…

After looking over what you have written about these three areas of your life, identify at least one goal for each area. These goals are helpful to keep in mind when you consider what situations you want to focus on in pushing through your anxiety and worries. You don't have to take on more than one goal at a time.

Often a situation will trigger your anxiety and it is easier to avoid the situation than to push through it. Setting a specific goal that you want to reach helps motivate you not to let anxiety and avoidance stop you from doing what matters to you. Remember that pushing through anxiety is the way you *change* how the amygdala responds so it produces less anxiety, allowing you to follow your goals rather than blocking them. Your life should be guided by your goals, not limited by your cautious amygdala.

17 Goal Setting: If You Didn't Have Anxiety, What Would You Do?

So many times, we have met clients who feel blocked by their anxiety. They want to live more fulfilling lives, but anxiety keeps them from pursuing activities and attending events that they want to participate in. The answer is goal setting. Think about what you would do if anxiety were not limiting your life. Focus on activities that you avoid but you would like to do if you could—like driving on the freeway, volunteering to give blood, speaking up when someone says something rude, asking someone out on a date, or riding on a Ferris wheel. The first step is just setting a goal that you would like to accomplish.

Goal setting needs to be approached correctly. First, it is essential that you focus on what *you* are interested in. Your goals should not be chosen by other people in your life, by what society tells you, or even by your therapist. Your goals need to come from your own interests and values. They should be a reflection of your hopes and dreams for your life. Focus on goals that will make a difference in your life, that will motivate and encourage you.

Second, be aware that anxiety sometimes keeps us from even thinking about certain goals. For example, your goal may be spending more time with friends, but if you get anxious eating in public, you may not feel comfortable thinking about spending more time with friends when it potentially involves going to a restaurant. When you think about spending time with friends, the amygdala now makes you feel anxious. This may lead you to abandon the goal before even taking it seriously because even thinking about the goal increases your anxiety.

Let's focus on some areas of your life that you may want to address. In the previous chapter you considered goals related to your daily life, your relationships, and your work life. Look back at what you wrote in your journal.

Rating Your Goals

After reviewing what you wrote, use your journal to make a list of some goals you are interested in. Consider it a wish list of what you would like to change. For each goal, complete the sentence, "I would like to..."

Then, rate each goal according to two different dimensions: Importance to You and Anxiety Level. Use a scale of 1–10, with the higher numbers reflecting more importance and more anxiety in the situation. See the example below.

Goals	Importance	Anxiety
I would like to go out to a concert with friends.	6	5
I would like to speak more at staff meetings at work.	8	9

Rating goals in this way can help you see their importance. When you work toward a goal, you want it to be something important. But you also want to consider how difficult a goal is for you, in terms of the level of anxiety you will be facing.

Working on goals that have higher importance and lower anxiety is a good way to begin. When you start out with situations that your amygdala does not respond as strongly to, you can get the benefit of accomplishing something important to you without confronting a high level of anxiety. You also learn from the experience. Yes, you experience anxiety, but you also see that you can change things. You see that when you put yourself in a situation that triggers your anxiety, you are giving your amygdala the opportunity to learn that the situation is not dangerous.

Although it's challenging to deliberately face a stressful situation, the amygdala learns and changes as a result of the experience. You feel anxiety as you work on the goal, but you also feel something very empowering: *Your anxiety comes down as you work on your goal.* When you feel this decrease in anxiety occur, you know that the amygdala is learning! You are rewiring your brain to produce less anxiety in that situation. You can actually *feel* this occur when you put yourself in the situation and work toward accomplishing your goal despite the anxiety.

18 Identifying Your Anxiety Triggers

The amygdala learns through association. When some kind of negative event occurs, the amygdala takes note of anything that is associated with the event. If you are in a car accident, whatever is associated with the accident, especially situations or objects that occurred or were present before the impact, are likely to become triggers for the amygdala. This means that, after the accident, your amygdala will react to icy roads, horns honking, the squeal of brakes, the intersection where the accident occurred, sitting in the passenger seat (if that is where you were), and perhaps even the song that was playing on the radio. The connection between the trigger and the negative experience does not have to be a logical one. Anything associated with the negative experience can become a trigger for fear and anxiety.

Consider some triggers that people experience as a result of the amygdala being watchful for anything associated with a negative experience or threat of some kind. A clown intended to be entertaining frightens a child and clowns terrify the child for years. During a bout of the flu, you eat corn on the cob (which you used to enjoy) and after this, even seeing it makes you feel nauseated. A woman is sexually threatened by a bald man and feels nervous around bald men for years. A boy only hears his middle name when he is in trouble ("Thomas Richard!") and eventually complains about how much he dislikes his middle name. An administrative assistant becomes anxious whenever she opens her email after experiencing stressful email interactions with her supervisor.

Triggers can be objects, sounds, smells, locations, experiences, and even bodily sensations. One of the authors even became triggered by the sensation of turning right in a car! After a car accident in which Catherine was hit while turning right, every time she turned right and felt that specific physical sensation of a right turn, her heart would start pounding and she'd get a rush of adrenaline. Her amygdala had associated right turns with danger.

A loud voice or an exasperated sigh can become triggers for people with a history of conflicts in relationships. Remember that the trigger doesn't have to be dangerous in any way; it only needs to have been *associated* with a negative experience. Just *pairing* the trigger with a negative experience is enough. A Vietnam War veteran started having panic attacks in the shower after his wife inadvertently bought the same soap he had used during combat. The harmless soap was his trigger.

Identify Your Triggers

What are the objects, situations, locations, sounds, smells, or tastes that are triggers for you? As you go through your day, be aware of when you feel the physical sensations that suggest the amygdala has been activated: a twinge of anxiety, a catch in your breathing, or even a strong jolt of adrenaline. Some triggers you may be well aware of, but, with some reflection, you may be able to recognize new triggers. Note some of your triggers in your journal, and allow yourself to really look at, listen to, or experience the trigger, observing what happens in your body and mind as your amygdala reacts. You are likely to react (or want to react) to the trigger in a way that reflects the defense response (fighting, fleeing, or freezing).

You may or may not be aware of how this specific situation or object became a trigger for you. But it is useful to recognize that the trigger is the result of the amygdala encoding an emotional connection of fear to this trigger. Very often, the trigger poses no danger, but experiencing the trigger does produce a very real emotional reaction that occurs automatically and without your control. That is the influence of your amygdala.

Begin to refer to them as triggers, for example, "Any mention of my weight is a trigger for me," or "His tone of voice triggered me." Keep in mind that the emotional reaction is something learned in your amygdala.

Identifying your triggers and recognizing the amygdala's role in creating your emotional and physical reactions is an essential step in taking back control of your life. While the amygdala-based reaction to your triggers is intended to be protective, often it restricts you from activities that are important to you. Recognizing triggers and understanding the amygdala underlies your reaction to them can help you overcome barriers that have kept you from living the life you want. It is really a game changer when you realize that a trigger does not necessarily mean danger and the emotions you feel are because the amygdala has *learned* to react this way. Your feelings are very real, but they don't mean a true danger exists. Even better, you can teach the amygdala to react in a new way and stop producing those feelings.

19 Talking Back to the Amygdala

Often, we give in to our anxiety to feel relief. We feel anxious or have worries, so we back out of plans we have made, or we stop trying to do something that we had hoped to do. While this provides temporary relief, it does nothing to help us manage these situations in the long run. In fact, now that we know that anxiety comes from our amygdala, we have learned that seeking relief from anxiety and backing away does nothing to change the amygdala. The amygdala learns nothing new about the situation when we give in to the anxiety. The amygdala will continue to produce anxiety about that situation every time we approach it, unless we teach it something different.

If you want lasting change, you need to change your amygdala. You can teach the amygdala to stop producing anxiety in specific situations, but first you have to change your thinking about your anxiety. You need to recognize that just because you feel anxiety does not mean a significant danger exists. In producing anxiety, the amygdala is sending you a message that certain situations are dangerous, and you need to back away. But your main evidence that a situation poses a threat is the feeling of anxiety the amygdala produces. Are you sure the amygdala is correct that the situation is as dangerous as the amygdala makes you feel? Can you imagine yourself handling the situation if the amygdala would just stop producing the anxiety that it brings every time you approach it?

Certain thoughts will also help you not give in to anxiety. You can think, *I want to be able to do things* and *I can resist giving in to anxiety* and *Backing away is not going to help me change things.* These thoughts may help you resist the tendency to seek relief rather than change. As Mark Twain noted, courage is not the absence of fear, it is acting *despite* fear.

A client came in for therapy fed up that she had allowed herself to get to the age of twenty without getting her driver's license. She had always felt anxious when she tried to drive and kept backing away from the experience. In therapy, she began working on pushing through her anxiety, and she agreed to just sit in the driver's seat in a safe area, knowing that she was going to experience anxiety just putting herself in that position. As she sat there, her anxiety came up to an eight out of ten and tears came to her eyes. She said, "I hate this feeling." The therapist said, "I'm sure it feels awful. The amygdala is throwing a fit. Just keep sitting here until the amygdala realizes that you are safe." In a matter of a few (perhaps ten) minutes, the young woman said, "I can feel my fear coming down. I thought this would take hours!" She began to talk excitedly, "This is what I have to do? Show the amygdala that it's safe by not leaving the situation?" After experiencing her anxiety decrease, she was very motivated to try more.

Talking Back to the Amygdala

You don't want to let the amygdala control your life, do you? When you realize that two little almond-sized parts of your brain are trying to run your life, you can start questioning the amygdala, rather than trusting the feelings that it produces. You can reject the amygdala's guidance and talk back! Here are some suggestions:

I am tired of you making me feel like this.

I am not giving in to you; I want to try this.

I want to set goals for myself.

Don't try to stop me; I can take at least these first steps.

You aren't going to scare me away from this.

I got this! You aren't stopping me this time.

Even though talking back to the amygdala won't change the amygdala, it can change you. You can plan your life instead of letting the amygdala dominate you. When you start to get frustrated and angry at how the amygdala blocks you, you can push past whatever feelings the amygdala brings up.

You can change the amygdala. By staying in a situation, you can teach the amygdala that the situation is not dangerous. You may be surprised that when you refuse to give in to anxiety, the amygdala backs down. The amygdala can learn to see a situation as safe when it is given time to do so.

20 Lessons for the Amygdala

The amygdala does not learn from coaching, lectures, or logical arguments. Nothing that we say to our clients during our sessions has a direct effect on the amygdala. As we have noted, the amygdala needs *experience* to learn. The amygdala produces fear, dread, anxiety, and panic, typically based on triggering events in your life. Whether you consciously remember the experience or not, the amygdala remembers. The good news is that you don't need to know how or when the amygdala learned to activate a response to a trigger in order to teach it to stop responding to that trigger.

To teach the amygdala, you need to seek out experiences that expose you to the trigger so the amygdala can learn something new about it and stop responding to it. You want the amygdala to learn that the trigger is safe. Note that not every trigger needs to be addressed. Some triggers can simply be avoided without impacting your life. But when anxiety about a trigger is getting in the way of something you want, like being able to drive through an intersection on the way to work, you'll need to provide some lessons for your amygdala.

No matter how old you are, the amygdala can still learn. The amygdala can acquire new fears and learn to stop responding to triggers. But to provide a lesson for the amygdala, you need to give it an experience to learn from. You need to put yourself in the presence of the trigger and show the amygdala that the trigger is not a threat. Of course, the amygdala won't learn instantly. In fact, the first reaction that the amygdala will have is to produce the defense response (the fight-or-flight response) and you will experience fear and anxiety in your body. The amygdala needs to be exposed to the trigger for a while before it learns to stop producing the reaction of anxiety to that trigger, but it will stop. After a few exposures to the trigger, the amygdala will learn to respond with calm instead of fear.

For the amygdala to learn the trigger is not a threat, no harm or negative events can occur while exposing the amygdala to the trigger. You must be in a safe situation and not in any danger. The problem is, until the amygdala begins to identify that the trigger is not a threat, the amygdala will produce anxiety, increase your heart rate, release adrenaline, etc. This makes teaching the amygdala challenging; even though you are not placing yourself in danger, you will likely feel anxious. It helps to recognize that just because you feel anxiety doesn't mean a danger exists. Remember how your amygdala can mistakenly make you withdraw in fear from a small dark shape in the shower that turns out to be a tangle of hair. In the same way, it can continue for years to react with fear to friendly dogs after your grandmother's dog knocked you down in enthusiastic affection when you were three. Exposing the amygdala to a trigger by providing experiences with dogs that are not associated with any harm allows the amygdala to learn that a dog is not a threat and stop reacting.

How long the amygdala takes to recognize that a specific trigger is not a threat can vary, but you can definitely feel when the amygdala stops producing the defense response and your anxiety decreases. Then you'll know you have just taught the amygdala something new. But how do you keep yourself near the trigger during that period of anxiety before the amygdala learns?

How to Encourage Yourself to Teach the Amygdala

These encouraging statements can help you stay in the presence of a trigger so that your amygdala will learn, even when the amygdala is producing anxiety. You might want to write them out in your journal to learn them, and jot down a situation to use each one in.

Now I've got the amygdala's attention!	*I can tell my amygdala does not like this!*
Go ahead and throw a fit. I'm staying put.	*You're not going to make me run away.*
I'm hanging in here until my amygdala learns.	*This is difficult, but I'm in charge here.*
I will feel it when my amygdala learns.	*I'm not going to let you run my life.*

These statements are useful whenever you are trying to push through anxiety, but they are especially helpful during exposure, when you are deliberately staying in the presence of a trigger. They help keep your cortex from activating the amygdala while you are trying to teach the amygdala that the situation is safe. The amygdala can't learn it is safe when you are thinking frightening thoughts.

21 Find Your Why

Have you ever played a sport, a musical instrument, or even a game in which you had to practice or try for hours? Maybe at some point you were exhausted and ready to quit but you kept striving. What kept you motivated to keep going? Did you dream about winning the championship or performing for others who applauded? Did you imagine yourself succeeding or your family cheering you on? Memories of why you trained and practiced so hard in the past, and images of what you achieved as you improved, can help motivate you to keep going when you face a challenge. When things get tough, we need motivators to keep going; we need ways to encourage ourselves to do difficult things.

When you are dealing with anxiety, change is not easy. Even when you want change, it can still be hard to work on something that is challenging. Maybe anxiety has been in your life for so long that it feels like a part of you. When you have gotten so accustomed to coping in specific ways, it is hard to think you could approach things differently, especially if you have struggled with anxiety for a long time.

In creating the defense response, the amygdala produces the very distressing feeling of anxiety, which pushes us to seek safety and comfort. It can feel counterintuitive not to obey this impulse. It is easy to believe that listening to your anxiety has kept you safe from harm, embarrassment, or pain, and that to push past your anxiety would mean you are opening yourself up to unpleasant emotions and maybe unpleasant experiences. Well, you'd be partially correct. When you are finally ready and willing to lean into anxiety, to stop avoiding and take some chances, you do have to tolerate some unpleasant feelings—at first. "At first" is important. The truth is that the amygdala will learn and change, but only if you take it into the very situations that it tries to make you avoid. Only if it experiences the situation will the amygdala learn it is safe and stop producing anxiety. You don't have to take on every situation, but more opportunities will be open to you once you see that you're stronger than you thought. You are capable of more than just avoiding, fleeing, freezing, or fighting (the limited options that your amygdala offers). As we noted, courage is not the absence of fear. Courage is *acting despite fear.* Taking the first few steps is hard because even if you have the courage, you don't yet have the experience of showing yourself that you can handle whatever comes your way when you disregard the warnings of your amygdala and move into the situation anyway.

So, when taking on a new challenge, it can be helpful to remind yourself of your *why.* Why are you doing this hard work? Why are you willing to ditch comfort for growth? Why is it worth finding the courage to make a difficult change? Why is this change important to you? Answering these questions can help you focus on what you stand to gain. Does your *why* have to do with improving your quality of life, relationships, career, parenting? Are others also likely to benefit?

To Find Your Why, Consider Your Whats

If answering the why is a little hard for you, try asking yourself the more concrete what? Take some time to answer these questions in your journal. What would I be doing if anxiety didn't hold me back? What could I be doing if I wasn't blocked by worry, apprehension, and fear? What could I be doing if I were acting instead of reacting?

What would be different if I chose courage instead of comfort and stopped allowing anxiety to dictate my behavior? What will make it worth it to push through anxiety? What goal am I moving toward?

These questions can help you think about your motivation for tackling anxiety and doing challenging things. You need motivation to find courage. Envision the changes you are seeking. Is your motivation related to relationships? Career? Quality of life? Travel? Something else? After you have written some ideas in your journal, jot down what you want on sticky notes, using phrases that start with "What I want is..." Put them around where you can easily see them each day, so that when facing anxiety feels challenging and you want to quit, you can be reminded of specific aspects of your why and keep going.

As you read this book and seek guidance for the path ahead, stop, pause, and think about what motivates you. What's *your* specific *why*? Find it, label it, nurture it, and own it. Post notes about *what* you want on your wall and look at them often—in these simple statements is the promise of a better life.

22 Don't Avoid Situations! Push Through the Fear

When anxiety happens, you naturally want to do things to avoid or escape it. Anxiety is designed to feel uncomfortable. Avoidance is a behavior that helps us feel better. It lets us escape anxiety in the short term but maintains anxiety in the long run. Avoidance prevents you from learning that the things you fear are not as dangerous as you think. Imagine you are afraid of dogs. You see your neighbor's dog. You may automatically run inside and when you do, you no longer feel anxious. But what happens in the long run? Avoidance maintains the anxiety because you never really learn that the dog is safe. Your amygdala may be tricking you about the dog being dangerous.

Avoidance feels good but doesn't allow the amygdala to learn. The trigger may actually be safe, but the only way the amygdala learns that is through experience. Also, avoidance strengthens the idea that all triggers should be feared. In order to reduce anxiety, we begin to avoid situations, people, places, things, and even triggering internal experiences like thoughts, feelings, and physical sensations. Avoidance can limit your life when you deny yourself experiences that are actually safe. We often tell people that avoidance masquerades as your friend, but it is actually your enemy. It allows anxiety to dominate your life. Avoidance is what the amygdala wants! But think about how avoidance has impacted your life. What has it cost you?

It's normal to believe that it's best to avoid anxiety-provoking things. Why would you seek out situations that are stressful? But, if you think about a situation or thing that you once feared but no longer do, you most likely overcame the fear by pushing through it. Maybe circumstances forced you to face a fear, such as when a child fears going to school but is required to go every week. Or maybe, someone encouraged you to face a fear, as when parents play in the water with their children, encouraging and praising them if they jump in or put their faces into the water. Over time, when you face fears repeatedly, you learn that what you fear may not be dangerous and you learn that you can handle it. Pushing through avoidance is one secret weapon for reducing anxiety in the long run. In the words of Robert Frost, "The best way out is always through."

When You Changed Your Fear

Think of a situation, object, or animal that you once feared, but you eventually overcame the fear by facing it. How did facing your fear help you change your feelings? Write about it in your journal. Did you have an experience in which you did not avoid the situation? Why did you have to remain in the situation? What led you to put yourself in that particular situation? Consider why you ended up facing this fear. Did someone help you? Did you push yourself to do it? Did it take a long time for the fear to subside, or did it change fairly quickly? Consider if, in this experience, you can see how facing fear is what changes fear, as long as the situation does not cause you any harm (other than the discomfort that fear and anxiety causes). Note thoughts that you have about these experiences in your journal.

Choosing to push through your fear feels unnatural so people can go years caught in the grip of a certain fear. The amygdala learns to stop producing the defense response, which causes you to feel fear and anxiety, only when the amygdala stays in the situation long enough to learn that negative experiences do not necessarily occur. (Well, feeling anxiety and fear is a negative experience, but no other negative experiences occur.) If you don't push through your fears, the amygdala has no opportunity to learn. Unfortunately, you can't change the amygdala by lecturing, explaining, or reasoning. It needs to have the opportunity to be in the presence of whatever triggers the anxiety and experience that nothing harmful or negative occurs. When this happens, you can actually feel the amygdala calm down. You often experience a feeling of decreased tension and relief that lets you know that the amygdala has changed in a very real way.

23 Facing Your Fears

If you want to reduce anxiety, your job is to confront your fears one step at a time and learn to tolerate uncomfortable feelings and thoughts. Your anxiety will decrease if you face your fear repeatedly and for a long enough time. The amygdala only learns through experiencing the trigger without having negative things happen. You teach the amygdala to respond differently by exposing yourself to the trigger, whether it is speaking in public, dealing with your child's vomit, or attending a crowded concert with friends.

The process of deliberately allowing yourself (and most importantly, your amygdala) to stay in the presence of a trigger is called *exposure*. When you engage in exposure, you may feel a moderate level of anxiety. However, if you stay in the situation and don't allow anxiety to make you retreat, you will find that after a period of time, your anxiety begins to go down, and you can actually feel this change. This helps you see that if you show your amygdala that the situation is not a danger to you, you can actually change the amygdala, so it stops producing the defense response.

Typically, we start exposure with easier situations and then try more challenging ones. Let's take the example of feeling afraid of a friend's dog. You could start by just being around the dog while your friend holds him on a leash. That would be your first exposure exercise. On another day, you could try moving closer and petting the dog while the friend holds the leash. Finally, another day you could try being around the dog when he is in the room and moving freely without a leash. In each situation, you need to be committed to not leaving the situation until your anxiety decreases.

What is happening is that you are giving your amygdala the opportunity to see that the dog is not a danger. The length of time that it takes for anxiety to reduce varies, but you will definitely feel a reduction if you give your amygdala time with a trigger when nothing negative occurs (except the experience of anxiety). With repeated and consistent facing of your fears, anxiety should subside, often more quickly than you expect. And if it takes some time for your anxiety to come down, that's okay too. You are learning that you can tolerate some anxiety, and it doesn't hurt you.

Face Your Fear with Exposure Exercises

Think about a trigger that causes you anxiety. Is your reaction to this trigger blocking you from doing something important to you? We only ask you to go through exposure for a good reason! Think of three activities related to this trigger that may make you feel at least a moderate level of anxiety. Write them down on a sheet of paper. Rate your anxiety for each activity on a scale of 0 to 100 (with 100 being an intolerable level of anxiety). Choose the activity that produces the least anxiety of the three and engage in that activity with the goal of not leaving the situation until your anxiety is reduced by half. If you rate the anxiety as an 80, you need to stay in the situation until it goes down to 40. After exposure, notice what you have learned. Did you feel less anxious fairly quickly, or did it take a while? Were you surprised that you could handle it? If your anxiety did not reduce by half, repeat the exercise until it does. If it did, repeat the exposure at least once or twice before moving on to the next activity. Remember the key here is to pick activities that create a moderate level of anxiety. If the activity does not create and then reduce anxiety, your amygdala isn't learning anything. Also, if you don't push through anxiety, you are not learning that you can do it and your worst fear is not likely to occur. Below are examples of three activities for someone who fears embarrassing themselves and thus avoids speaking to others. Remember to stay in the situation until your fear goes down by half.

1. Walk around in a mall, making eye contact and giving a half smile to others. (30)
2. Go to a family gathering and engage in conversations with relatives. (60)
3. Initiate conversation with a stranger by asking questions about them (e.g., how was your day?). (90)

Carrying out exposure exercises is challenging, but it is very important to do them correctly. Leaving the situation before your anxiety goes down by at least half can actually cause anxiety about the trigger to increase because it strengthens rather than decreases the amygdala's reaction. Don't engage in exposure exercises on your own unless you are certain that you can get yourself to stay in the situation until that point. We strongly recommend that you work with a skilled therapist who is trained to use exposure correctly. Therapists know how to avoid common problems in exposure, so if it doesn't work, consult one! This is an amazingly powerful intervention, if it is done correctly.

24 That Silly Amygdala!

The amygdala has been shaped by our prehistoric past, and as a result, its patterns of responding are best understood when you look at how it could have been helpful to our ancestors. One of the essential purposes of the amygdala is to keep us safe from danger. The amygdala developed to be watchful for predators and remember whatever was unsafe—whether it was a location, an object, a sound, or even a taste or smell. Anything associated with danger was stored in memory in a way to identify it as unsafe. So, when you approach a location or even smell a scent connected with a bad experience, the amygdala produces the defense response and a feeling (like fear, anxiety, or dread) that indicates danger.

While this was useful to our ancestors, it's less helpful in our twenty-first-century world. For example, it makes sense to fear a location where you encountered a tiger and thus avoid that location and a possible tiger attack. But does it make as much sense for a child to be afraid to go to school after a bad thunderstorm occurred while they were in the classroom? That is the kind of emotional learning that the amygdala produces.

Fears produced by the amygdala can result in some strange reactions that only make sense when you understand how the amygdala learns from something being paired with a bad experience—whether or not the location or object that is feared had anything to do with the bad experience itself. A person can feel an intense fear when hearing a specific Beatles song just because the song was playing in the car just before an accident occurred.

A college student might feel a rush of adrenaline and a pounding heart as they enter the classroom to take a Psychology exam and wonder how a response designed to help them run away from danger is helpful. In our modern world, we rarely face dangerous predators, but we are still living with brains and bodies that prepare us to run away from them. The amygdala can react to a bill collector on the phone like a tiger in the jungle!

Isn't the Amygdala Silly?

Taking the amygdala less seriously can help us get less caught up in the emotions it produces. Read through the following experiences our clients have had, and then see if you can come up with some ways in which you can call out your own amygdala.

When she realized that spiders really do not pose a danger, even though her amygdala made her feel like running away, one client said, "C'mon amygdala! I'm taking you down in the basement with the spiders whether you like it or not!"

A CEO of a company figured out that her nervousness about being around her coworkers drinking and laughing at parties was related to her childhood memories of her drunken and belligerent father who had long ago overcome his drinking problems. She said, "Honestly, my amygdala is acting like I'm still a five-year-old! I'm the damn CEO!"

When a socially anxious teenage girl learned there would be a family reunion at her home, she panicked and asked her therapist what to do. The therapist suggested that she could jog around the neighborhood during the reunion to reduce her anxiety. The teenager rolled her eyes but tried it. After the reunion, she reported, "I was so stressed. I told Mom you said to go for a run." She was stunned that when she came back home she was calm and able to interact with her relatives. "I even talked with my uncles! My anxiety was gone! What happened?" When told that the amygdala produces the fight-or-flight response and running was exactly what the amygdala wanted, she said, "But I ran right back home. Geez, the amygdala must be stupid!"

Now it is your turn. Get out your journal and under the title My Silly Amygdala, write down some examples of when your amygdala seems rather silly. Start your examples with phrases like these:

I can't believe that my amygdala thinks...

My amygdala makes no sense when it...

Apparently, my amygdala thinks it is dangerous to...

My amygdala wants me to run away from...

My amygdala would like me to get into a fight with...

Learn to take the amygdala less seriously, and recognize that the reactions it produces are, on a certain level, misguided and out of date. Many times, it misleads us into being afraid of something that poses little danger or makes us feel full of unnecessary fury.

25 Identifying Values: What Is Important to You?

Sometimes we are so dominated by fear and anxiety that we lose sight of what is important in our lives. We focus on trying to find relief and our lives are dominated by the goal of keeping ourselves from experiencing anxiety. We lose touch with our goals and values because of this narrow focus. We have difficulty seeing the big picture. In this chapter, we will invite you to clarify your values for yourself and identify what is truly important, in order to allow those aspirations to become part of your life again.

Values are different than goals. They are beliefs about what is important in life and what expectations you have for yourself in your different roles in life, in your job, in your family, in your community, and on the planet. We don't achieve a value like we do a goal. Instead, our values are the beliefs that influence what goals we select, and they motivate us to achieve our goals so that we feel more satisfied with who we are.

The pressures of life can keep you busy with tasks and responsibilities that may not clearly relate to your values. In addition, anxiety can distract you from your values and dominate your goals. You may neglect situations that are important to you as a result of anxiety or focus so much on keeping your anxiety under control that you have no time for acting outside a very narrow sphere. When you take time to get in touch with your values again, you may be able to choose goals that are more aligned with who you are and who you want to be.

Getting Back in Touch with Your Values

Consider each of the areas below to see if specific areas relate to values that you have in your life. Are there personal qualities that you want to bring to each area? If so, write down at least one value that you would like to follow in that area of your life. What values would you follow if nothing was stopping you? You don't need to identify a specific value for each area.

Spiritual: Begin at the broadest level by considering what spiritual and religious traditions resonate with you. Take time to identify some values from these traditions that you admire and respect. Do specific values resonate with you?

Physical Health: Do you have strong values relating to maintaining your physical health?

Relationships: What are your values in relationships? These values may vary, depending on the nature of the relationship. Consider the types of relationships that are important to you. Romantic partnership? Friendship? Co-workers? Answer the following questions for any relationships you consider important in your life. What do you value in terms of what you want to bring to your relationships? What do you value in what you receive from relationships? These words may help identify what you value in relationships: Love, Patience, Power, Adventure, Calmness, Learning, Freedom, Fun, Safety, Fairness, Happiness, Harmony, Honesty, Humor, Involvement, Learning, Loyalty, Independence, Playfulness, Respect, Reputation, Stability, Reason, Wealth.

Family: What values do you have for yourself in your various family relationships?

Education: Do you value education as a part of your life? If so, would you like to pursue more education?

Work and Career: What are your values in your career or at the workplace? How do you want to approach your work?

Experiential Values: What types of experiences do you appreciate? Do you feel it is important to have music, nature, dance, sports, recreation, humor, contemplation, delicious food, or other experiences as part of your life?

Taking time to connect yourself to your strongest values can help you to decide how to set goals for yourself. It can also clarify why you may not feel invested in certain activities, or why you don't feel satisfied about certain aspects of your life.

You may notice that you have gotten out of touch with a value that means a great deal to you. You may also notice that some of your values conflict with each other, and you find you have to choose which is more important. Difficulties like these are common, and if you are aware of them, you can make more conscious decisions about what is important in your life.

26 Welcome Fear (It's a Game Changer)

When you pay too much attention to the bodily sensations and emotions that the amygdala creates, the amygdala can become a bully that bosses you around. For example, if you think, *What if I get cancer and die*, and then the amygdala produces anxiety as if you are truly in danger (when it is just reacting to a thought), you are likely to react to that anxiety. In order to calm that anxiety, you start doing research about cancer symptoms, checking for bumps on your body, and trying to reassure yourself that you do not have cancer. These strategies may reduce anxiety in the short term but serve to feed the anxiety in the long run. You devote more time and attention to the thought and increase your anxiety because you react to the anxiety created by the amygdala as if it were something you need to get rid of.

We talked about facing fear and anxiety as one strategy for reducing anxiety in the long term. Another strategy is to go a step further and *welcome* fear. For example, if we imagine the amygdala as a bully, we may say, *Hey, it's you again, trying to scare me*, and then simply continue with whatever you are doing. You can think, *I am welcoming this opportunity to push through fear to teach the amygdala. I'm not backing away*. Similar to welcoming the fear is thanking the amygdala. In *The Happiness Trap* (2008), Dr. Russ Harris describes an exercise called "thanking your mind." Try remembering your brain is engineered to keep you safe and thank your amygdala when it tries to scare you.

Instead of engaging with the thoughts your mind creates, thank your amygdala for being so concerned and trying to protect you: *Thank you, amygdala, for trying to protect me; Thank you, cortex, for having such creative thoughts!* But you don't have to trust your thoughts or anxiety. We very often have no evidence for a thought, but when the amygdala produces anxiety, we suddenly feel it should be taken seriously. This is not necessary.

Recognizing that creative thinking about potential dangers is not always helpful and knowing the amygdala reacts to thoughts in the cortex as if they *are* reality, we can thank the amygdala for its concern and cautionary anxiety, but also say, *I got this, amygdala. Appreciate your concern, but I'm focused on other stuff right now.* We can do this with whatever thoughts make us anxious. By doing this, you are showing your amygdala that you are not afraid of something just because the amygdala apparently is.

Welcoming Fear and Anxiety

Take one anxious thought that comes to mind and try welcoming it, and the anxiety that the amygdala produces in response, with a neutral, open stance. You can also try to welcome it with humor. In your journal, write out statements that help you welcome fear like "Thank you for reminding me I may develop cancer. I know you are trying to protect me with this anxiety. Why don't you come along with me while I continue on this project I was completing for work? I appreciate you!" When you do this, it's also helpful to take note of your physical response. Are you cowering or grimacing? You'll want to welcome the anxiety with an open and neutral demeanor such that your physical stance and facial expressions do not show a scared reaction. Relax your body, hold up your head, and smile. When bullies get this response, they are likely to go pick on someone else.

Write down what you notice in your journal. Is this a completely different approach for you? Is it hard not to get caught up in focusing on the thought? Did your anxiety decrease faster when you tried to welcome anxiety?

Welcoming anxiety may sound counterintuitive and be quite challenging. Remember that to have courage means to acknowledge fear is there, take a breath, and move forward *with* the fear. Courage is acting *despite* fear not without it.

27 Resisting Compulsions and Safety Behaviors

What tends to happen when you feel anxious? You probably engage in some behavior to reduce anxiety because it's uncomfortable and distressing. All sorts of behaviors serve to reduce anxiety. Some of these behaviors are called *compulsions* or *safety behaviors*. A compulsion or safety behavior is any behavior performed to prevent a fear from coming true, and although it reduces anxiety in the short term it only maintains anxiety in the long run.

When you engage in these behaviors, you are not learning that you can tolerate anxiety and face your fear. You miss a chance to teach the amygdala that what you fear is not likely to occur. Instead, you begin to believe that you need to do the behavior every time you feel anxious. Also, because the feared outcome did not occur when you engaged in the compulsion or safety behavior, you give that behavior credit for the relief, when relief can be achieved in other ways.

Say you get triggered by the worry, *What if I have a heart attack?* That thought activates the amygdala, and you feel anxiety. To reduce anxiety, you check your heart rate (a safety behavior). Your heart rate appears normal, so you feel relief, but then you have the thought again, and you need to check again and again such that now you have a compulsion, and you are checking many times a day. You also can mistakenly believe that you kept yourself from having a heart attack because you checked your heart rate, instead of learning that you don't need to focus on the worry at all. Common compulsions include scanning or checking your body for signs or symptoms, searching on Google, washing or cleaning, seeking reassurance, going to the doctor excessively, avoiding activities, or sitting near an exit.

You can't always see a person's compulsions. Mental compulsions are behaviors that you do in your mind to reduce anxiety, like mentally reviewing and analyzing an event over and over or providing yourself reassurance repeatedly. Compulsions and safety behaviors strengthen amygdala activation because the temporary relief you get is like a reward the amygdala gets for producing anxiety. The amygdala has succeeded in getting you to do something and temporary relief followed. The amygdala wins and anxiety continues...

Strategies to Reduce or Eliminate Compulsions and Safety Behaviors

1. Track the compulsion by recording the number of times you engage in it. Research shows that self-monitoring can help you change a behavior (Balaghi et al. 2022).

2. Try to reduce the frequency of the behavior or eliminate some part of it. For example, if you tend to examine your skin and feel for bumps, only allow yourself to do one of these.

3. Delay using the behavior. Even if your urge is strong, you can tell yourself that you can resist for the next fifteen or thirty minutes and do it after that. Building your ability to resist gives the amygdala time to learn. Notice how you become more able to resist!

4. Engage in a valued act instead of the behavior. For example, go for a walk, call a friend, or take a hot shower instead of engaging in the compulsion.

5. If the behavior is mental, you can resist by focusing your attention on watching a game, making a phone call, or counting backward from 100 by 7s. Because these tasks require all of your attention, it is difficult to engage in a mental compulsion at the same time.

6. Remind yourself why you don't want to engage in the compulsion. These could be things like, *I don't want the amygdala to control my life. I want to have less anxiety. I don't want to waste my time on this anymore.*

7. Try to sit with the anxiety without doing anything, even for just a few minutes. This may feel uncomfortable at first but you'll find that it should get easier the more you practice.

8. Engage in progressive muscle relaxation or mindfulness instead. Focus your attention on being curious about your experience of anxiety.

9. Get help from someone. Ask a trusted friend or your significant other to engage in a pleasurable activity with you when they see you using a compulsion.

Compulsions and safety behaviors can seem to be helpful in the short term but notice how in the long term they keep the amygdala producing anxiety. To stop the amygdala from repeatedly producing anxiety in this particular situation, you need to reduce these behaviors and tolerate the thought or situation so that the amygdala learns the situation is safe. Otherwise, these behaviors will prevent the amygdala from learning that it is not necessary to produce the defense response in this situation.

28 Opposite Action

The sensations and feelings of anxiety are the most powerful tools the amygdala has to influence you. When you experience the activation of your sympathetic nervous system and the accompanying feelings of fear and dread, you can feel that every fiber of your being is telling you that you are in danger. You feel you must run away, stop trying, or fight against whatever is happening. At first, what you are experiencing is not on the level of thoughts; you are experiencing very primal and basic emotional and physiological reactions intended to keep you safe. But in our modern world, these reactions are often not a reflection of true danger.

Nevertheless, when you experience these reactions, it is natural to interpret them as indicating actual danger. So often, the worst, most catastrophic thoughts come into your mind when you feel this way and add fuel to the fire. Now you not only *feel* the drive to avoid situations, people, places, or things that seem dangerous, but you begin to *believe* that avoiding them is necessary. On one level, you may realize that there is not a real threat, but your bodily reaction is a powerful influence. The truth is that when you escape, you are usually just avoiding the feelings of dread and anxiety produced by your amygdala, not any real danger.

We get such relief when we back away from situations in which we experience anxiety! It feels so much better when the amygdala stops producing those terrible sensations. Unfortunately, you don't always notice how your world gets smaller and smaller due to avoidance and anxiety. The relief you seek is putting limits on you and taking away your freedom. Avoidance causes you to lose control of your life as you stay inside the boundaries set by the amygdala. When the amygdala protests *Don't do it!*, you comply. This maintains anxiety because the amygdala does not have a chance to learn the situation is safe by actually experiencing the situation.

Opposite action is an important skill to help you manage anxiety in a way that prevents the amygdala from limiting your life. Opposite action comes from dialectical behavior therapy (DBT; Linehan 2014) and essentially means you do the opposite of whatever the amygdala is telling you. The opposite action involves *doing* what you are afraid of. This begins by simply *approaching* what you are afraid of, which may seem a bit easier. Approaching what you fear will help give you a sense of confidence and mastery over the anxiety. Essentially you are responding to your anxiety as if it could be tricking you, and not assuming there is danger. Also, you are showing the amygdala who is in charge! Opposite action for anxiety entails these steps:

1. Identify the feelings of bodily activation and anxiety, as well as the "urge to avoid" that is associated with these feelings, and recognize them simply as sensations and feelings, which is different than proof that the situation is dangerous.

2. Determine whether the urge to avoid would be healthy and helpful in this situation or just a response that maintains anxiety and limits you.
3. If the urge to avoid is not helpful or healthy, engage in opposite action: approach instead of avoid.

Trying Opposite Action

Think about something that you want to do but are avoiding because of your anxiety. Would engaging in the activity be healthy and of value to you? Choose something that would be of value to you. Now, choose the opposite action. Take a step toward approaching the activity.

For example, if you are feeling very anxious about going to a party and speaking with people you don't know, you may feel the urge to avoid the party. Avoiding the party reduces your anxiety in the short term but may leave you socially isolated and without the chance to show your amygdala that speaking to new people is not a danger. The opposite action here could mean recognizing your anxiety but going to the party anyway for a short time. You may expect your anxiety to stick around, but it will likely decrease once you are in the situation. The amygdala creates anticipatory anxiety but may not stay activated when in the actual situation. By engaging in opposite action, you are not letting the amygdala dictate your actions and behavior.

Remember that the amygdala is often mistaken about whether a situation is dangerous. Also, remember that the amygdala learns from experience. Athlete Walter Anderson notes, “Nothing diminishes anxiety faster than action.”

29 The Miracle of Mindfulness

Have you ever driven home from work and been surprised when you made it home because you were just "zoned out?" You didn't notice the trees or the roads because you were lost in your thoughts. We are often not paying attention to the present moment. And when we are paying attention to the present, we are busy judging, analyzing, and evaluating it. Have you ever heard the phrase, "Stop and smell the roses?" Imagine keeping your thoughts focused on the here and now, enjoying the smell of roses and the birds flying in the sky.

Mindfulness is the act of intentionally bringing your attention to the present in a nonjudgmental way with a focus on acceptance (Orsillo et al. 2004). The goal of mindfulness is to use your senses to notice and observe your internal experiences and the world around you without judging. For example, you can use your senses to taste a piece of candy or notice all the different cars as they go by. You can also notice the sensations in your body or the thoughts that you are having without judging them. You can observe your thoughts like you are watching clouds pass by in the sky or leaves float down a stream.

Simply observing thoughts helps you not to get entangled with their content. It gives you distance from them, allowing you to see them like a bystander rather than getting wrapped up in them. This can help anchor you in the present and create calm because you are simply observing and noticing what is happening in your mind instead of believing a thought or reacting to it as bad or harmful. A thought can just be a thought versus something to be feared. Taking a thought too seriously activates the amygdala, and it is easy to get entangled in repetitive worry thoughts about the meaning of the thought when it is simply a thought.

Mindfulness helps to calm the amygdala by redirecting attention and focus to the present, which is typically safer than the worried thoughts we often focus on. When you practice mindfulness, you get better at choosing what you focus on rather than letting your anxiety control your focus. We know the human brain is easily distracted and constantly shifting focus. If you have a tendency to think, *What if this or that bad thing happens*, mindfulness can help you to notice those thoughts but redirect your attention to the present moment. For example, *I notice I am having the thought that my partner no longer loves me, but I am choosing to continue to observe and notice the trees and flowers, the sounds of birds chirping, and the warmth of the hot summer sun on my skin*. We benefit from choosing what to focus attention on because not every thought deserves all the attention we give it.

Mindfulness is a great skill because we always have our senses and can always rely on them to anchor us in the present moment, rather than get caught up in thoughts that just produce distress for no reason. Because our brains have the tendency to analyze, judge, and worry, we need to practice mindfulness in the same way we strengthen a muscle.

Turn to mindfulness when you notice yourself focusing your thoughts on situations that distress you. Refocus yourself on something as simple as your surroundings. What do you hear, see, and smell? This pulls you into the present. You don't have to let your mind wander wherever it wants. You can grab the steering wheel and take charge with mindfulness. Focusing on your breath is another way to return yourself to the present moment and away from amygdala-activating thoughts.

Practicing the Skill of Mindfulness

One simple exercise is to practice noticing and observing your breath. Give yourself a few (4 to 5) minutes to practice this exercise. Once you have experience, you can extend the time. Stop what you are doing and sit quietly. Now direct your attention to your breath. Notice if it is shallow or deep, fast or slow, warm or cool. Feel it in your chest, your nose, your mouth. Count your breaths. Your brain may wander to other thoughts or judge your breath; this is normal, and when this happens, simply observe it, and bring your attention back to your breath and the simple observation of it. You may need to redirect yourself many times and that is normal because the human mind is a wanderer. Notice your mind wandering and bring it back to the breath each time. Take some notes in your journal. What did you notice? Was it difficult? How often did your brain wander? Did you feel a sense of quiet or calm when focusing on your breath?

Our busy thoughts can distress us; often they are focused on judgment and evaluation of the past or the future. Other times, they are focused on worries about the future that may never come true. Mindfulness is a great skill that can help us stay present-focused with simple acceptance and observation. As you practice it, you strengthen your ability to be in charge of your focus. You can choose what you focus on and take more control of your life!

part three

Calming the Cortex

30 What Starts in Your Cortex Doesn't Stay in Your Cortex

When you have difficulties with anxiety, it helps to know how the brain produces it. You know the amygdala is the part of the brain that produces the defense response, activating the sympathetic nervous system and creating the bodily sensations that we experience as fear and anxiety. Anxiety *cannot* be produced directly by the thinking part of the brain, the cortex. Only the amygdala can make all the bodily changes we experience as fear and anxiety. The cortex is capable of so many amazing things; the cortex allows us to interpret what we see and hear, have thoughts, hold memories, carry on a conversation with ourselves in our head, and use logic and reasoning. The cortex can influence whether we experience anxiety, but not in the way that you may think. Understanding the connections between the amygdala and the cortex in your brain will help you learn new ways to control your anxiety.

The cortex has relatively few connections to the amygdala, compared to the amygdala's many connections to the cortex. Therefore, the cortex has only a few ways to monitor or influence the amygdala, while the amygdala can monitor and influence the cortex in a variety of ways. The cortex does not monitor all processes occurring in the brain. The cortex only monitors and shares the *results* of what we see; for example, it does not monitor how cells on the back of the eyeball collect light waves and send the information through the optic nerve to other parts of the brain for processing.

It is the same way with the amygdala. The cortex does not monitor or report back to us what is happening in the amygdala, so we are not conscious of what happens in the amygdala, any more than we are conscious of exactly what happens in our eyes and brain to produce vision. We are only made aware of the effects of the amygdala when the cortex allows us to experience changes in our bodily sensations like increases in muscle tension or heart rate, or the feeling of dread associated with anxiety.

In contrast, the amygdala's connections to the cortex allow it to be informed about what is happening in the cortex. When the cortex produces its detailed analysis of what a person is seeing and hearing, the amygdala is able to monitor the sensations, perceptions, and thoughts that are occurring in the cortex. This means that what you are seeing, hearing, and thinking does not simply stay in the cortex. The amygdala watches what goes on in the cortex, like watching a television screen. This also means the amygdala can respond to the thoughts produced in the cortex whether they are true or false.

For example, if your sister frowns when you say that your child will be attending a certain school, and (in your cortex) you have the thought that she doesn't approve of the school and is judging you, your amygdala is monitoring that thought. The amygdala may produce a small burst of anxiety as a result. Your sister may be frowning because she is trying to remember where that particular school is located or is wondering whether your child will be walking or riding a bus. But if you are thinking that your sister is judging you harshly, your amygdala reacts to that thought.

In the same way, if Nick checks his phone and sees that his girlfriend has not texted him, he may have the thought that she is going to break up with him. His amygdala may produce a surge of anxiety in response to this thought even though he has no evidence that his thought is correct.

Observe the Influence of Your Cortex

During the next twenty-four hours or so, observe your thoughts to see if you can detect times when a mere thought can cause a reaction in your amygdala. At first, you may have difficulty noticing the thought before you experience an anxious feeling. It may be that the first thing you notice is a feeling of anxiety, and you may need to pause and consider what you were just thinking. Increasing your awareness of how thoughts in your cortex influence your amygdala will give you more ways to manage your anxiety. Write some observations in your journal.

Amygdala-activating thoughts are a common problem for people trying to cope with frequent anxiety. Becoming more aware of these thoughts can help you recognize where the anxiety begins. If you can interrupt and change your thoughts, you'll most likely experience less anxiety.

31 Change the Channel!

Sometimes, we are filled with anxiety because troubling thoughts keep going through our minds. Whether we are thinking about bills to pay, problems in relationships, or stresses due to work, we often know it's not healthy to dwell on these thoughts. You can feel the defense response build as the amygdala reacts to thoughts like these. You become filled with anxiety, but it is as if you can't look away from these thoughts. Some of us are very good at getting trapped in this thinking.

Ruminating is the word for lengthy periods of negative, repetitive thoughts, typically about something that poses a potential problem or threat. The word rumination is also used to describe the process of repetitive chewing seen in cows, as in the expression "a cow chews the cud." When we ruminate, we are chewing on certain troublesome thoughts, over and over, and not moving on to other thoughts. Rumination has been linked to clinical depression, as well as generalized anxiety disorder and OCD.

We mentioned that the amygdala monitors the thoughts we have in the cortex and may react to these thoughts as though they are true events, not just thoughts about events that may occur. In a way, we can think of the amygdala as watching cortex television, just sitting there watching whatever distressing images and ideas are being created and elaborated on in your cortex. Unlike a television, however, the cortex has no "off" button. Instead, you have to change the channel!

If you understand how the cortex works, you will be more able to change the channel. To get rid of a thought, you can't just tell yourself to stop thinking it. Ironically, if we ask you right now to not think about pink elephants, pink elephants will immediately come into your mind. That's because by bringing up the idea of pink elephants we have activated the circuitry in your cortex that stores information about pink elephants, and there is no way that activating this circuitry about pink elephants will help you stop thinking about them. Instead, the best way to stop thinking about pink elephants is to *replace* the thoughts with something else. If I ask you to imagine a huge tortoise with purple hearts painted on its large shell and a rose in its mouth...the idea of pink elephants leaves your mind.

In the cortex, you can't erase a thought—you must replace it with something else. Telling yourself not to think about the idea or image that you want to stop thinking about just activates the very circuitry you want to stop activating. Telling yourself, *Don't focus on that overdraft fee right now,* activates that circuitry and keeps the overdraft fee in your mind. Turning your focus to something else is like changing the channel. Focusing on something else will be more effective in making sure you are tuned in to a channel that will not activate the amygdala and cause you anxiety.

Practice Changing the Channel

During the next twenty-four hours, make a special effort to be aware of when your thinking activates your amygdala, leading you to experience anxiety. Use the concept of changing the channel in your cortex. If your television has hundreds of channels, your cortex has millions of channels. With practice, you can change channels relatively easily!

Changing the channel means focusing your cortex on something different than the amygdala-activating thoughts that you want to replace. New thoughts could be about training the dog or the mystery novel you are reading. Thoughts can involve doing something, like singing a song or talking to someone. You can plan something for the future or remember something from the past. The replacement thoughts just need to hold your attention. They can involve fun activities like playing a game on your phone or calling an old friend to catch up. They can focus on a task in your daily life: planning your grocery list or making the employee work schedule you need to send out. For ideas, answer these questions in your journal:

> *What do I want to think about? What do I need to think about?*

Remember that we aren't asking you to change the channel because certain thoughts are dangerous. They are merely thoughts. We want to change them because they have a negative effect on the amygdala and cause unnecessary anxiety in your life.

32 Don't Trust Your Cortex

People tend to believe that the cortex allows them to experience reality. But that is not necessarily true. Our cortex does not make a video recording of what is happening, nor does it make sure we attend to all the sounds around us. In fact, the cortex is very capable of providing us with incorrect information. When you recognize that the amygdala reacts to what is going on in your cortex, and often relies on the cortex to determine whether you should worry about something, you realize that some skepticism about the cortex is helpful.

First, let's make sure that you recognize that the cortex is not always correct. Read the following:

Once upon a a time there was a wizard

If you aren't careful, your cortex will remove something that is very clearly there. Can you see that your cortex missed a word?

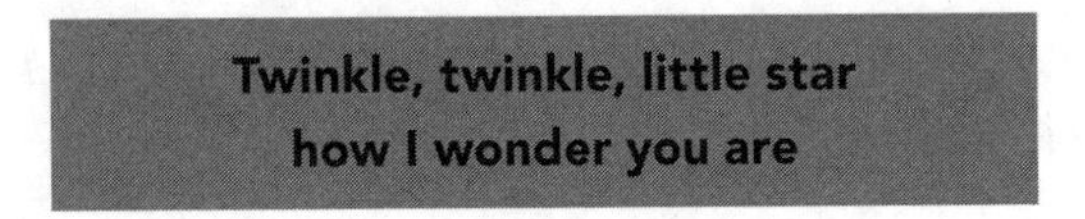

Your cortex can also add information that is not there. Did your cortex insert a word for you in the box above? The point is that the cortex does not carefully reproduce the reality around you. The cortex *creates* an experience of reality, and it actively manipulates the information that it processes. What you see is your cortex's *interpretation* of reality, not reality itself. And this is not only true of vision, but also of other senses and perceptions of reality. The cortex actively creates our perceptions, tidying up what is processed based on what is expected or assumed. Just take a look at the following sentence, which is nonsense, but which your cortex can quickly make sense of:

Yuo cna rade tihs setnenec eevn touhgh jsut oen wrod is spleled croretcly.

The plain truth is that, at times, the cortex can make you see things that are not there. Sometimes a friend is trying to assure you that she is not angry at you, but you can't stop worrying that she is. Or perhaps you have been reassured by a doctor that you don't have cancer, but you keep finding bumps or spots on your skin that you are concerned are cancer. You can also fail to see something that is obviously there, but other people can see clearly. A friend may go out of her way to do kind things for you, but you still doubt that she really cares about you. Or you do a presentation at work that your boss says was great, but you just don't believe that it was good enough. The cortex is perfectly capable of making incorrect interpretations, and we need to be careful not to trust it.

Do You Trust Your Cortex Too Much?

Consider these questions and write down any examples that you can think of.

1. *Do you often have thoughts or worries come into your head that you feel you must take seriously, even though you have little or no evidence to support that they are true?*

2. *Do you tend to believe that you know what people are thinking, when they have not told you their thoughts?*

3. *Do you often worry about something occurring that never comes to pass?*

4. *Do you find yourself unable to believe people when they try to explain the reason why they did something, and assume that your assumptions about them are correct?*

5. *Have you assumed that someone was disappointed in you when in fact they were not?*

6. *Are you very good at imagining how future situations could go badly, and tend to take your imagined outcomes too seriously?*

7. *Do you dwell on the meaning of a thought for an hour, when it is just a thought?*

If you answer yes, and can easily think of examples, in response to many of these questions, you have a tendency to treat your expectations, anticipations, assumptions, and worries with more attention and respect than they deserve. As a result, your cortex is creating a great deal of unnecessary anxiety—because these are just the kind of thoughts that activate the amygdala.

What we think is not always accurate. Our brains do not capture reality as it exists, but rather create an interpretation of reality, based on previous experience, assumptions, emotions, and many other factors. The problem is not just that *you* might believe the thoughts; the amygdala monitors what is happening in the cortex and may also react by creating the defense response. The amygdala will often react to thoughts as if they are true, or as if something you worry about will occur, which causes you to experience anxiety. Just remember—when anxiety strikes because of the thoughts you are having, don't assume you should take it seriously. You can't always trust your cortex!

33 Defusion Skills

The truth is that most people have horrible, unwanted thoughts, images, or urges. But most people don't think too much about these thoughts and don't take them seriously. In contrast, many people with anxiety difficulties become preoccupied with distressing thoughts and may even believe that if they are having the thought, image, or urge, they may end up acting on it. We call this *thought-action fusion*. For example, imagine having the thought as you are driving down the road, *What if I drive my car into oncoming traffic?* Almost everyone has this thought, but most can dismiss it as just a thought. Those with anxiety are more likely to have thought-action fusion, meaning they believe that because they had the thought, they may act on it. This belief—that they will act on a thought—can create a great sense of anxiety and fear.

Thought-action fusion makes it difficult to dismiss the thought because the person believes they need to pay attention to it and perhaps do something to prevent themselves from acting on it. Thoughts are just thoughts. Reacting to them and dwelling on them makes them stronger and more intrusive and generates more anxiety. So, as you can imagine, thought-action fusion leads to more problems.

In addition to thought-action fusion is simply *fusion*. Fusion is when we believe that our thoughts always mean something; we believe they need our full attention and we need to be very concerned about the scary ones. For example, if you are having the thought *I am worthless and horrible at my job,* in a state of fusion, you believe having the thought means it could be the absolute truth. But thoughts are not truth. If thoughts were true, we could simply think we are going to win the lottery and make it happen. But thoughts don't really have that power. In the same way, just because we worry that someone is late because they were involved in a car accident, the worry does not make an accident happen.

One skill that can help you with fusion and thought-action fusion is a technique called *defusion*. Defusion is the act of deliberately creating distance or separation from your thoughts. One way to defuse thoughts is to put the stem, *I am having the thought that*...before stating the thought you are having. *I am having the thought that my daughter was in a car accident.* You can further defuse a thought by adding another layer to the stem: *I am noticing I am having the thought that...* This helps you look at the thought as an observer instead of getting tangled up in the content of the thought. See how different it is when you say, *I am noticing that I am having the thought that I am worthless and horrible at my job?* Using these phrases changes the way you relate to the thought. You don't have to believe the thought or act on it but simply notice that it is occurring and you're aware of it.

Practice Defusion

Think of a thought that creates anxiety. Pick a thought that frequently recurs and is upsetting to you. For example, *nobody at work likes me*. Think this thought for a few seconds. Notice the feelings that arise as you focus on the thought. Write down what happens. You have gotten the amygdala's attention, haven't you?

Now put the phrase *I am having the thought* in front of it. What is it like to say to yourself, *I am having the thought that nobody likes me*. Note if you see a difference in your reaction.

Then put the phrase *I notice I am having the thought* in front of it. So, *I notice I am having the thought that nobody likes me*. Note any difference in your reaction.

You probably noticed that adding these stems gives you some distance from the thought. You are turning it into a thought you are observing, instead of fusing with the content of the thought. Making this change in the cortex can help you see that, if you focus on controlling what is going on in your cortex, you have more control over your amygdala's reaction.

In *The Happiness Trap*, Russ Harris explains that when we use defusion, "thoughts may or may not be true. We don't automatically believe them. Thoughts may or may not be important. We only pay attention if they are helpful" (2008, 41). He adds that we do not have to obey thoughts, or follow their advice, and notes thoughts themselves are not threatening. Try defusing your thoughts and notice if they feel less believable or threatening to you!

34 Use Gratitude to Combat Anxiety

According to the Oxford dictionary, *gratitude* is the quality of being thankful, a sense of appreciation. Focusing on gratitude helps to shift not only your mood but also your perspective. Practicing gratitude does not mean denying that you experience pain and suffering; it means frequently choosing to focus on appreciating moments of beauty, joy, or affection. You are taking control of the pathway that your thought processes follow and not just allowing your brain to wander according to its own tendencies. The human brain is programmed to be preoccupied with potential difficulties, but we can work to resist this programming.

The amygdala can influence the cortex to focus attention on any threat or potential threat that the amygdala is responding to. For example, you recently had a fight with your partner and thoughts about the fight keep coming up, even when you are driving or trying to focus on your work. The amygdala wants you to stay alert to any danger to the relationship and can hijack your cortex to focus on this fight. The cortex can even get further involved and by starting to analyze in detail what the fight means. As you feel the anxiety build (due to more negative thoughts in your cortex, which activate the amygdala even more), you begin to interpret the anxiety you are feeling as indicating that the fight means more than you initially thought. You begin to consider whether you will break up, and anxiety builds further as your worry circuits start to consider all the what-ifs and catastrophic possibilities.

We have discussed this negative spiral; your cortex is hijacked by the amygdala to increase focus in on a threat which further activates the amygdala. We noted that one way to avoid this negative spiral is to focus on staying in the present, the here and now. Mindfulness is one way to focus on the present. Another strategy for staying in the present is to focus on what we are grateful for instead of what we are worried about. Practicing *radical gratitude* is training yourself to experience gratitude even in your worst moments of anxiety or suffering.

The goal of radical gratitude is not to be happy, although you may feel more happiness. The goal is to appreciate that beauty, meaning, and kindness are present along with pain and suffering. When we chose to focus on any positives we can find in our lives and practice gratitude, our brains release dopamine (Burton 2020), serotonin (Pu-Linck et al. 2007), and oxytocin (Burton 2020)—chemicals that help us feel better. If you choose to direct your attention toward positive ideas and what you are grateful for, you reduce amygdala activation. Whatever thinking patterns you direct your attention to will become stronger, so you are actually rewiring your brain's tendencies. Research has shown that practicing gratitude decreases anxiety and increases hope, life satisfaction, and positive moods (McCullough et al. 2002).

Ways to Increase Your Focus on Gratitude

Start each day by identifying three things that you are grateful for, three things that went well the day before, or three things that you are looking forward to. Be more specific than *I am grateful for my wife.* Try *I am grateful for my wife who always greets me when I arrive home and lets me know she loves me and is happy to see me.*

End each day by using the acronym GLAD to review the day for four things. Identify one thing you are Grateful for, one thing you were glad to have Learned, one thing that you Accomplished, and one thing that Delighted you (Altman 2014).

Keep a gratitude journal by getting a notebook and taking time at the beginning of each day to jot down what you come up with from one of the two activities above.

Set an alert on your phone for several times during the day, and when it goes off, come up with something you are grateful for in the current moment or location.

Express thanks despite having difficult times. Make a point to give thanks to others whom you appreciate or for a job well done.

Choose one or two of these gratitude practices. Working gratitude into your daily life can gradually help decrease your anxiety. You are reducing the brain's natural tendency to focus mainly on potential threats that activate the amygdala. Often it helps to involve someone else in this practice, and both agree to report to each other what you are grateful for each day. Perhaps exchange GLAD reflections with each other via phone calls or texts. This helps you make a commitment to regularly including gratitude in your life. The very act of noticing and describing positive aspects of our lives and what we are grateful for is an effective skill in fighting off anxiety. You will be surprised how quickly you can turn your mood around when you make a deliberate effort to look for positives in your life.

35 Put Your Cortex in the Courtroom

Now that you know the amygdala reacts strongly to thoughts that are catastrophic, you don't want to be tricked by the amygdala and automatically believe that the anxiety you feel in response to a simple thought is justified. Anxiety can make you *feel* like something bad is going to happen, when all you are really feeling is the amygdala reacting to a thought. The feeling of anxiety can make you believe that the likelihood of some negative event happening is very high, but the amygdala cannot accurately predict the future. It is simply trying to prepare you to deal with a *potential* danger, which in many cases never occurs.

The feeling of anxiety can distort your thoughts and make them seem more compelling and believable. Anxiety can lead you to make the mistake of thinking that your thoughts are foretelling the future, creating further anxiety. Instead, take a moment to identify the thought you're having so you can evaluate or modify it. Bring some logic into the situation. The process of evaluating and modifying your thoughts is called *cognitive restructuring*. Whenever you are trying to adapt or change thoughts, you are trying to modify the cortex.

The first step is to identify the *situation* that led to your anxiety. For example, maybe your spouse is not home on time. Next, identify the *thoughts* you are having about this situation that are creating anxiety. Perhaps you think they were in a car accident or something harmful occurred. Then try to identify any cognitive distortion you may be engaging in. In this case, it is catastrophic thinking. You are assuming that some terrible, catastrophic thing has happened. Notice that the *thought* that something bad happened is what activated the amygdala which produced anxiety. The situation of your spouse being late did not cause the anxiety by itself, it was the thought of a car accident. People often jump to unrealistic conclusions because of pessimism, catastrophizing, or mind reading (thinking you know what someone else is thinking). Then the amygdala reacts to the unrealistic thought and creates anxiety.

Before believing a thought, shouldn't you determine if it's accurate or not? We tell people to imagine they are in a court of law. What evidence do you have to present to defend this thought? And what evidence is there that it may not be accurate? In this example, we don't really have any real evidence for the thought that something disastrous has happened. No one has called you to say something bad has happened. You can come up with evidence against the thought. Your spouse has been late before for just mundane reasons. There are alternative explanations like they stopped at the store, their boss called them into a meeting, or they met a friend after work. Given all the evidence for and against the thought, is there clear evidence that your thought is true? Or is the feeling of anxiety tricking you into thinking there is a danger because you have a *feeling* of danger? Do you want to trust an emotional reaction to a thought that has no proof? Focus on what is realistic or believable. You may come up with an alternative thought such as *Maybe it's just that they are tied up at work.*

You can see that by replacing your thought, anxiety may still be there, but lessened. In examining the evidence, also remember all the times you felt anxious and that many times the feeling of anxiety did not accurately predict a problem. Remind yourself, *I have felt anxious many times, and most of the time my thoughts were not accurate, so I don't want to put myself through anxiety until I have evidence that my thoughts are accurate.* Just like in a court of law, when a person is presumed innocent until there is proof of guilt, assume a situation is not a danger unless you have evidence. We tell people, if the situation can have many explanations, why believe the one that is going to make you feel the worst? Situations are not always in your control, but you can use cognitive restructuring to adapt your thoughts so that you don't experience distress until evidence shows you there is a reason to be distressed.

Cognitive Restructuring to Combat Amygdala-Igniting Thoughts

Write this pattern in your journal:

Situation → Amygdala-Igniting Thought → Anxiety

Choose a situation in which you experience(d) anxiety to write into the pattern.

Draw an arrow and write an amygdala-igniting thought(s) that come to mind.

Follow the pattern, drawing the arrow to Anxiety. That is where the thought will lead.

Consider the evidence for and against your thought(s).

Consider several alternative thoughts to replace the thought.

Come up with an adapted thought that has more evidence.

Notice if you feel less anxious after you have restructured your amygdala-igniting thoughts.

You may think thoughts are automatic, frequent, and difficult to challenge or adapt. This is true. However, if you practice this skill daily—even if it's not in the moment—you can begin to challenge thoughts, even as they are occurring, more effectively. You can begin to rewire your anxious brain and gain more control over your anxiety, instead of letting it control you.

36 Using Your Worry Circuits Correctly

Worry is a uniquely human ability. Other animals don't worry, by which we mean they don't actively contemplate possible negative events that could occur in the future. But our worries do not predict the future in any reliable, accurate way. Often, we worry about events that never happen and experience negative events we never anticipated.

Worry must be useful for humans, or it would not have developed in the human brain and be so common. The circuitry that produces worries developed in the frontal lobes of the cortex, behind our forehead and eyes. Being able to anticipate potential negative events before they occur could be helpful—especially if our ancestors responded to worries by *making plans* that allowed them to avoid or prevent the negative event or to cope with it in some way.

Imagine a prehistoric woman sees a pack of wolves while gathering berries. When she returned home, she imagined the pack of wolves might harm her infant if she left the infant unattended. She was worried about a terrible event that had not happened. But her worrying would only be helpful if she made a plan to prevent this negative event from occurring. Luckily, in the frontal lobes of humans is circuitry for planning, which allowed our ancestors to come up with step-by-step plans. This can be seen in how humans have devised methods of hunting, building structures, and planting crops.

As you observe your own tendency to worry, consider whether you are using your worry in a helpful way. Worry is most helpful when it alerts you to a potential concern and allows you to make a plan to remedy or cope with that concern (Wilson 2016). If you worry but don't make a plan, it would be as if the prehistoric woman kept dwelling on the danger of wolves but never considered keeping her infant in a safe place. Not only is worry by itself a useless exercise, but the amygdala reacts to these thoughts and images and you experience anxiety.

If we are going to experience worry, we want to make sure that we don't suffer the anxiety without gaining some benefit! The benefit is in the plan. Only if the prehistoric woman examines her shelter and tries to make it secure against wolves to protect her infant, has she benefitted from worry. After worry alerts her to a danger, she needs to move from worry to making a plan.

Evaluating Your Use of Worry

The beneficial use of worry can be illustrated in this simple diagram.

Worry → Plan →

Write in your journal about any tendencies you have to fall into these common traps:

Stuck in Worry: When you worry (i.e., think about potential negative events that could occur in a certain situation), do you tend to keep worrying, by thinking about various different things that could go wrong, and imagining these negative outcomes, without thinking about coming up with a plan to deal with it? You just keep cycling in the worry and don't move to the plan? This leads to the amygdala producing feelings of anxiety with nothing gained. Make a plan and then move on.

Come Up with a Plan, but Go Back to Worry: When you worry, do you realize that you need a plan to prevent the situation from occurring or to cope with the situation, and make up a plan, but then go back to worrying? In this case you are soaking in amygdala-activating thoughts when it isn't necessary. Move on, not back to worry.

Come Up with Multiple Plans, but Always Find a Flaw in the Plan: This approach is basically worrying about every plan you make. If you find a flaw, come up with a better plan, but if you keep doing this, you are basically worrying about every plan you make. You don't need a Perfect Plan. You may never even use it, if your worry does not come true. Choose a plan and move on with your day.

Making a plan is what helps you move on from worry. And you don't always have to carry out the plan, if the need for it doesn't arise. For example, Greg is worried that Deja is irritated with him for telling a certain joke and keeps feeling stressed about this. If he comes up with a plan for how to handle the situation, it allows him to stop worrying. His plan is that he is prepared to make an apology, but he won't give it if she doesn't seem irritated at him the next time he sees her. He is ready with a plan if needed, but it may turn out that he doesn't need it. You can see how having the plan helps Greg end his worrying.

37 Relief in the Palm of Your Hand

The brain can do a million things at once: keep your heart beating, help you see around you, send hunger signals or turn them off, etc. But when it comes to *focusing attention*, your brain can only focus on one thing at a time. There is no such thing as multi-tasking when it comes to focus. You can rapidly shift your focus of attention from one thing to another, but your brain *cannot* focus on more than one thing in a given moment. It's just impossible.

This limitation of the brain is a precious gift to those of us whose anxiety arises from the thinking processes in our cortex. If your amygdala is activated by worries, obsessions, mind-reading other people's thoughts, or other cortex-based thought processes, you can minimize your anxiety simply by changing your focus.

Imagine that the limits of your focus are illustrated by the palm of your hand. Put out your hand, palm up, and think about how much you could balance on that palm. It's a limited space, and while it could hold a coffee cup or a cell phone, it can't manage too much at once. So let's say it represents the one thing you can focus on. That limited space in the palm of your hand is a very small area, and you are going to take charge of your mind by deciding what to focus on.

Whether you want to stop obsessing, worrying, or focusing on what other people are thinking, you can accomplish it simply by changing your focus. You just need to seize that small space allowed for your attention, and direct it to something else, and you have won a battle. It is a very small piece of real estate that you are fighting over. You can win this!

You can take charge of your focus in many ways, so choose the one that works best for you. Here are some suggestions. Any of these are better than thoughts that activate your amygdala!

A Change of Focus

- Whatever you are doing, do it mindfully. If you are eating spaghetti, focus on the smell, flavor, and texture of the food as you eat. If you are washing dishes, tune into the experience and feel the warm water and the dishes you are holding, searching for remnants of food to wash away. If you are listening to music, focus on the experience and sing along.

- Or, make a plan about something that you need to do, or are looking forward to. This could be as simple and mundane as making a shopping list for the week, or as complicated as googling international locations for a potential vacation. You could also plan your garden for the spring, considering what you want to plant. Think of a meal or dessert that you want to prepare and consider when and for whom you want to prepare it. You can look into consumer ratings of an appliance or auto that you are considering purchasing. Think of how you are going to entertain your grandchildren the next time you see them or start to think through a project at work that you need to accomplish soon to anticipate how you will handle it.

- Reach out to a friend you need or want to touch base with via phone or text. Or make that call you have been putting off and schedule an appointment or get some needed information. Just ask yourself, what do I need to do that would be good to get accomplished right now? This may be the right time to clean out the cat litter or take the dog for a walk.
- For stubborn thoughts that won't go away, the most effective approach is to have someone talk to you. You know how hard it is to focus when someone is talking!? So turn on the television or radio, listen to a podcast, or call a friend. Just make sure that whatever you focus on is not anxiety-provoking. Don't call your friend to talk about your worries, for example. Instead, talk about planning an activity you can look forward to.

Any time that you take charge of the focus of your attention, and make it a productive or pleasurable thought process, rather than focusing on thoughts that increase anxiety, *you win!*

38 Get Off the Anxiety Channel!

If your cortex is like a television that your amygdala watches, then the Anxiety Channel is one of the amygdala's favorite channels. The amygdala is designed to focus on the aspects of your life most important to survival, and so it is constantly on the lookout for threats—in your environment, in your relationships, in your future. The amygdala can influence the focus of the cortex, encouraging it to tune into any signs of threat. Your cortex can become trained not only to focus on threats, but to anticipate and watch for anything that even suggests a threat. The anxiety channel is when the cortex is in a state of vigilance and a habit of focusing on potential indicators of threats.

The way habits develop in the cortex is through practice. The more you think about a certain situation the more you activate circuitry storing certain memories and ideas, and the more likely those thoughts become. Schwartz and Begley (2003) refer to this process in the cortex as "Survival of the Busiest." The more frequently you focus on certain thoughts, the more you activate that circuitry, and the stronger that circuitry becomes, almost like wearing down a pathway. Pretty soon those thoughts become so strong that they dominate your thinking. And because they keep popping into your mind, you begin to assume they must be important, or they wouldn't be coming up. You can get stuck on the Anxiety Channel, worrying about the meaning of a frown on your friend's face, anticipating financial problems, or thinking you have mice in the kitchen cupboard...

We are not meant to be constantly focused on potential threats, but we often find ourselves stuck on the Anxiety Channel. Often the most intelligent and imaginative individuals are best at tuning into and strengthening their focus on this channel because they can use their logic and creative abilities to make it a very compelling, convincing, and even engaging channel to focus on. Our current 24/7 access to the media can feed into the tendency to focus on the negative and potential threats. But it can cause us to overestimate potential difficulties, imagine dangers that never occur, and (most importantly) activate the amygdala to produce anxiety when it isn't necessary.

Getting Off the Anxiety Channel

If you notice that you often catch yourself on the Anxiety Channel, you need to take a proactive approach. Take some time to think of new topics and activities to focus on. What else could your brilliant and creative cortex focus on? It could be recreational topics or activities, or it could be work related... You could plan activities or remember a past happy event that you cherish. Your quality of life is at stake here.

You may need to give yourself permission to take your focus off the Anxiety Channel. You may be assuming that it is productive or even protective to stay focused on potential threats. But if you find your focus being dominated by worries and anticipating potential problems, you are living through worse emotions than are necessary. The amount of stress you experience in your life can be reduced just by changing your focus.

Increase your time focused on the Play Channel. Give yourself permission to play—plan a fun activity, play a game on your phone, read something entertaining or amusing. Entertaining activities engage your cortex without activating the amygdala to produce the defense response, so your body is able to stay more relaxed. You also are more likely to smile or laugh.

You can start by making a list of anything that you can think of on the Play Channel. Having a list can really help you to refer to when you are wanting to shift your focus. You can also refer to it to see how successful you are at making use of the different ways to entertain yourself.

Focus on the Gratitude Channel at least once a day. Ask yourself to consider people in your life you are grateful for and think of the reasons you value them. Think of where you experience beauty, delight, or wonder in your life, and allow those experiences to be your focus for a while. Remember the deer you saw in your backyard, the fresh tomatoes your neighbor gave you, the silly tumbling of your grandchildren.

Even focusing on a task for work or a household chore can provide a channel that helps you to calm the amygdala. Taking time to focus on a concrete task and get something accomplished can be more productive and less stressful than the Anxiety Channel. Do you ever notice that when your life is particularly busy, you feel less anxiety? It's better because you don't have time for the Anxiety Channel.

When you want to switch off the Anxiety Channel, remember just how many different ways you can focus on something else. So many different channels are in your brain! You can count by 7s backward from 100, practice mindfulness by focusing on different senses, count all the blue things in the room, call a friend...and it is so helpful to have a list of ways to Play!

39 Replacing Worry with Problem-Solving

Think about all the times you have actively worried and worried about something. Thoughts like *What if I lose my job; how will I survive; what if I can't provide for myself,* and *What if I can't find another job,* etc. Your cortex can conjure up so many what-ifs! Worry is a behavior that can maintain and worsen anxiety. When we worry, we have thoughts involving catastrophes and arguments and all kinds of possible problems. Worry itself is not productive because it does not prepare us for what could go wrong or help us solve a problem. We focus on all the possibilities of things that could go wrong but that have not happened yet. We feel more anxious by dwelling on all the worst-case scenarios because the thoughts and images activate the amygdala. Instead of worrying, problem-solving is a skill that can help you reduce anxiety because it helps you devise a plan for dealing with the potential situations you are focused on.

Before beginning problem-solving, you need to ask yourself if there is a real problem to solve. Is it something you have control over? Is it an unsolvable worry or one with a very low likelihood of happening? If the problem you are worried about is out of your control, then it is not a problem that can be changed by action on your part. If you are worried about the outcome of an election, or about your neighbors selling their home and moving, you have no control over these situations. You can only focus on how to respond to the situation, not how to problem-solve and change it. However, if it is a real problem in the here and now that you can do something about, then using problem-solving can be helpful. Here are some examples of solvable worries: My rent is due and I may not have enough money. I have many household tasks to complete and am overwhelmed. My child is acting out when it is time to do homework. In contrast, unsolvable or very low probability worries are things like, *What if I get cancer in the future? What if my spouse gets in a car accident? What if a tornado destroys my home?* Can you see how some worries are not situations under your control or ones that have a low probability of occurring?

Problem-solving entails several steps in this order:

1. Identifying the problem that you wish to tackle.
2. Listing all possible solutions to the problem you can think of.
3. Evaluating each of the solutions by coming up with pros and cons.
4. Choosing the solution you think best to implement.
5. Implementing the solution if currently needed, or saving the plan for when it is needed.

After the chosen solution is implemented, you can step back and evaluate how helpful the solution was. If the solution is helpful, then great. If it is not, then you can go back and identify and choose another solution to implement, and so on.

An Example of Problem-Solving

Let's go through this process for a specific worry. Imagine that you have to pay rent soon and you do not have enough money. First, stop thinking about all the possible what-ifs, and then...focus on these steps to problem-solve!

Step 1: The problem to tackle is paying rent in a few days.

Step 2: Some possible solutions include: getting a second job; asking a family member for money; asking your landlord for some extra time to pay; trying to sell something; asking for an advance at work; moving in with a friend or family; and seeing a financial advisor who may be able to help you.

Step 3: Come up with pros and cons for each solution. For example, pros and cons for getting a second job. Pros: you could make money. Cons: you won't have any free time and it doesn't help you in the immediate moment. You can go over pros and cons for each possible solution listed above, as if paying your rent were a problem in your life.

Step 4: Based on the evaluation, which of the solutions above would you be most likely to choose? For example, you could choose to ask a family member to borrow money because a family member has savings and would allow you to pay it back gradually...

Step 5: Implement the chosen solution: I was able to pay the rent this month using the solution.

Now that you have seen an example, choose a problem that you are dealing with. Get your journal and go through the steps, writing down your ideas for each step.

The next time you catch yourself dwelling on what-ifs, remember these steps and try problem-solving instead of worrying. See if it helps reduce your anxiety by reducing the time your cortex is focused on the stressful what-ifs. Shift to making plans, and, if needed, taking steps to solve the problem. How does it feel to try this approach?

40 Radical Acceptance

When anxiety is experienced, it is natural for most people to want to find ways to reduce it or avoid it. Unfortunately, the saying "what you resist, persists" is true for anxiety. When you struggle with something or just want the problem to go away, it tends to get worse. Think about anxiety like a swing. What happens when you push a swing hard to try to get rid of it? It just comes back and will keep coming back. What about when you don't push it away? Eventually it slows down and stops. In a similar way, if you just allow anxiety to be, it will be more likely to stop. This is why we have discussed facing fears and leaning into whatever creates anxiety as much more effective strategies to manage anxiety in your life.

Radical acceptance is a skill that can help you reduce resistance and ultimately reduce anxiety. Radical acceptance comes from dialectical behavior therapy (DBT; Linehan 2014) and it focuses on accepting anxiety, instead of resisting it. Once you accept anxiety, anxiety is likely to reduce. Radical acceptance means accepting "it is what it is," essentially accepting reality as it is and not trying to fight or resist it. For example, if you have panic symptoms like trouble breathing or increased heart rate and you do not have any medical issues according to a doctor, radical acceptance would be to accept that you have these symptoms and not try to fight them or explain them or get rid of them. They will happen from time to time, and life goes on.

Radical acceptance is making a conscious effort to honor situations as they are instead of avoiding, resisting, or wishing the situation were different. This is a completely new perspective for many because we are so often trying to control and modify situations to fit our preferences. Radical acceptance is challenging when it comes to anxiety because anxiety is designed to be unpleasant. We are naturally wired to find it uncomfortable, and we typically avoid it whenever we can. We resist and try to minimize it in any way possible. The idea of accepting anxiety can be particularly challenging for people to grasp, but it is also very freeing if you master it.

Approaching anxiety with acceptance means that we allow ourselves to experience anxiety without trying to control it and to simply experience it without trying to judge it or resist it. To be clear, radical acceptance does not mean we want anxiety, like it, approve of it, or are inviting it. It just means we are accepting the situation as it is. The following activity can help you practice this approach to anxiety.

Practicing Radical Acceptance of Anxiety

Think about a time you felt anxious, and you did not have any control over the situation. Consider that part of your anxiety and frustration was about resisting and trying to be in control. Fighting the feelings creates additional suffering and prolongs your suffering, according to this perspective. Can you see how this happened in that situation?

The next time you face anxiety, try to practice radical acceptance, using the following steps:

1. Recognize that when faced with an anxiety-producing situation, you have a choice to accept the feelings and sensations that occur in the present moment or to fight them. Give up fighting the feelings and sensations. They are here. Just accept them.

2. Self-monitor your experience. Notice your reactions and allow yourself to experience them. Where do you feel your anxiety? What happens when you just allow the sensations and feelings to be and don't engage in a struggle against them? What happens when you admit that you don't have control, but that is something you can accept? It is what it is.

3. Use coping statements to help you manage your response to the situation. Examples include:

    - "Stressing over things I cannot change will not help me."
    - "I can accept hard and challenging situations; it's only temporary."
    - "Fighting pain and distress only creates more suffering; it is what it is."
    - "I can accept things the way they are."

How is this experience different than the typical way you resist anxiety? When you don't try to manage or control the experience of anxiety, do you focus on the anxiety more or less? How did you feel when you experienced anxiety under these conditions?

According to Buddhist philosophy, pain plus resistance equals suffering. Pain is a natural part of life. Learning to accept it rather than fight it will help reduce anxiety and suffering.

41 Show Yourself Some Compassion

No doubt you have heard that critical inner voice tell you, *You are not going to succeed. You can't do it; you aren't good enough*. We all hear that voice at one point or another. Those of us with anxiety may hear it more often than not. Recognizing and labeling those critical thoughts for what they are is an important strategy for reducing anxiety. Critical inner voices activate the amygdala's defense response and produce anxiety. When critical inner voices become a continuing habit, life can be very stressful. *Self-compassion* is a skill that can help you quiet that inner critic and learn to speak kindly to yourself. Self-compassion turns compassion inward and quiets the stress response. Compassion is the feeling we have when we witness someone's suffering and desire to help. According to leading expert on self-compassion Kristen Neff (2022, 2), "from a Buddhist perspective, compassion is omni-directional and includes oneself as well as others." When we extend compassion to ourselves during times that we are experiencing suffering or perceived inadequacy, self-compassion is composed of three elements: self-kindness, humanity, and mindfulness (Neff 2022).

Self-kindness means being non-judgmental and understanding toward ourselves when we suffer or fail rather than being critical. We are often kind toward others when they suffer or fail but are not as willing to be kind to ourselves. Self-kindness means actively showing concern for ourselves as we deal with adversity. We should treat ourselves at least as well as we treat others, don't you think?

Suffering and feeling inadequate is part of the shared human experience; the fact that you experience hard times and suffer means you are human. We all encounter stresses, losses, things that don't go our way, as well as make mistakes and fall short. Therefore, accepting this as a normal part of life rather than judging yourself allows you to experience more compassion for yourself. Also practice mindfulness, which involves being in the present moment without judgement and observing your thoughts and feelings as they come, without getting tangled up in them.

To respond in a self-compassionate way toward yourself after having the critical thought *I'm such a failure*, you may say, *I'm noticing that I am having a difficult time right now*; *I recognize that we all struggle to get things right. It's part of being human, and I am going to show kindness to myself.* Think about how you would console a friend. Would you make the same critical comments to them as you do to yourself? Sometimes, in order to act with self-compassion, it helps to ask, *How would I speak to my best friend?* Then, act that way toward yourself. If you can do it at least sometimes, you are off to a great start!

Practicing Self-Compassion

Below are four ways to practice self-compassion. Choose one or two to try by writing some ideas down in your journal. If you find this difficult to allow yourself to do, you need to recognize that you are too identified with those critical inner voices.

1. *Externalize your inner critic.* Think back to a time where you were very critical of yourself. Now imagine that voice as a cartoon character, what would they look like and how would they sound? When that voice pops up, you can begin to distance yourself from those criticisms by identifying that it's that character being critical and not you and you can work toward quieting it. You can practice talking back to that inner critic. What would you say to that character or person if they were speaking like that to your friend? Write down how you could argue back to support yourself against this criticism.

2. *Write a self-compassion letter to yourself when you are having a hard time.* Imagine that you are speaking to yourself like you would to a cherished friend. Be accepting and kind. How would you reassure that friend that they are worthwhile and valuable? Write down some statements like, "You are valuable, and you are loved. I believe in you. You are strong, and you will be ok."

3. *Physical touch.* If you are having a hard time, you can place one arm around your opposite shoulder and press firmly as though you are giving yourself a hug or touching the shoulder of a friend. Notice the warmth of your own hand and how soothing this can be versus being critical. Pat yourself gently and add a self-compassionate statement, "Wow, this is tough; we all struggle."

4. *Practice mindfulness.* Simply observe and identify your thoughts and feelings, listen to them, and allow them to be there without judgment or trying to change them. Remember that they are thoughts, and you don't have to take them to heart. Nurture yourself with statements of self-compassion. Allow yourself to complain about these thoughts or express regret, as well as be critical of yourself. But remember these are just thoughts and you deserve compassion for thoughts that cause you suffering.

Many of us feel very awkward about being compassionate toward ourselves and are more comfortable being critical and punishing ourselves. We need to consider if those thoughts are helpful. They tend to wear us down, not strengthen us to face our challenges. Do we treat ourselves the way that we would treat a friend? Can you be more compassionate and understanding toward yourself? What kind of statements can you make toward yourself that help give you peace?

part four

Resisting Cortex Traps

42 Thinking Traps Trigger the Amygdala

Your mind generates thousands of thoughts each day. Some thoughts are helpful and accurate, and some are not. Just because a thought occurs does not mean it is meaningful or even worth attending to. Remember, because the amygdala constantly monitors your thoughts, thoughts can activate the amygdala. The amygdala is always waiting for some threat to occur and biased toward preparing you for the worst that could happen. That is the amygdala's job: to prepare us for danger with the fight-or-flight response. If you think about something threatening, the amygdala prepares your body to respond. However, when the amygdala reacts to thoughts about threats that are not certain, it's like a fire alarm going off when there is no fire.

Understanding that your amygdala monitors the thoughts in your cortex helps you to see that it makes sense that if you are having scary thoughts, you will feel anxious. The amygdala is reacting to your thoughts. People who struggle with anxiety tend to believe that because they are thinking about something, there must be some truth to the thought. In addition, they feel anxious, as the amygdala ramps up the defense response, which seems to provide confirmation that a threat exists. It helps to be aware of common ways that people get stuck in thinking processes that worsen their anxiety. Then you can work to avoid these kinds of thoughts. We call these *cognitive distortions* or thinking traps. We all get stuck in certain patterns of thinking, so don't feel bad if you realize you often get stuck in one of these thinking traps.

All-or-Nothing Thinking

All-or-nothing thinking is a very common thinking trap. This involves seeing things in black and white categories or extremes without any shades of gray or any in-between. Something is either good or bad, success or failure, perfect or absolutely awful. People who think in all-or-nothing terms usually classify themselves and their experiences more negatively than people who see things more complexly. Any mistake will make the entire project a failure. Someone who receives a B on a test is unable to see the positives of what they accomplished, and the grade is just a disappointment. Someone who can see shades of gray in the world can see ways in which they did well on most of the test and identify the small parts they faltered on. All-or-nothing thinking leads to thoughts that activate the amygdala because so many situations are classified negatively. We feel a sense of dread when we focus on the negative.

Combatting All-or-Nothing Thinking

- When you think in absolutes and see situations as always one way or another, with no in-between, then you are an all-or-nothing thinker. If you notice that you are doing all-or-nothing thinking, keep in mind that few things are absolutely one way or another; in fact, there are usually even more than two dimensions to almost any situation. A useful trick is to replace the "or" with "and." An experience can be both good and bad! For example, instead of saying "I did horrible because I was given suggestions for improvement," you could say "I got some suggestions for improvement on my evaluation, and the majority of comments were positive."

- Another way to combat all-or-nothing thinking is to challenge yourself to see more complexity in the situation. If you go to a concert with a friend, and you find the performance to be a disappointment, instead of saying, "That was a waste of my time; the concert was awful," you could think of more dimensions, like saying which parts of the concert were better than others, and perhaps note that the best part of the evening was the drive to the theatre, and spending time with your friend. Finding more than one dimension or emotional reaction to the event can be helpful.

From this one example of a cognitive distortion, you can see that your cortex may be prone to generate thoughts that activate the amygdala and keep us stuck in anxiety. A first step in reducing the grip of anxiety is to identify these distortions, especially when they make more you likely to focus on the negative, even when the whole situation is not negative. When you are feeling anxious or worried in a situation, take some time to review your thoughts, writing them down if it helps, and see if you can identify any cognitive distortions like all-or-nothing thinking. Then see if you can come up with a healthier thought!

43 Mind Reading

Have you ever been afraid to speak in a group or ask a question because you are worried that others will judge you or think negative things about you? Do you ever interpret a frown from a friend to mean that they had a critical thought about you? This is very common and happens to many of us. We call this *mind reading* because we often assume we can tell what others are thinking. The truth is we are not really reading the person's mind, but simply imagining that we know what the person is thinking. How likely is it that we truly know what someone else is thinking? If you guessed "not likely," you are correct! It is very difficult to know exactly what someone else is thinking. Very often, the person's thoughts are not even focused on you or whatever you assume they are thinking about.

When we have concerns or worries about what others are thinking, those thoughts come from our brains, not from theirs. Thoughts like *They think I'm stupid* or *They are irritated with me* are very common *cognitive distortions*. These thoughts reflect our own self-doubts and concerns more than they accurately identify what others are thinking. And when you are imagining you know what others are thinking, your amygdala reacts to your ideas in much the same way it would if you had actually heard someone say, "You are so stupid." When the amygdala reacts to your thoughts, creating the defense response, you feel anxiety, and it makes it easy to believe that the person's thoughts are something to be concerned about, that others are critical of you, etc. Mind reading makes it difficult to approach a situation because your thoughts are filled with ideas and images that activate the amygdala and increase your anxiety.

Consider the Effectiveness and Effects of Mind Reading

First, how about trying an experiment? Grab a friend and ask them to think about something. Now try and guess what they are thinking. Check to see if your idea is correct. You aren't very good at mind reading, are you? That's because it is very difficult to know what someone else is thinking. For example, it is common to assume a person is thinking about you when their thoughts are on a completely different topic.

Next, identify some situations in which you feel at risk to use mind reading, creating more anxiety for yourself when you don't even know what the person is thinking. In your journal, identify three or more situations in which you realize you have been mind reading in the past.

Write about the impact of your mind reading in these situations. Do you ever feel even more worry or anxiety when you assume that you know what others think, or will think? Does your mind reading ever cause you to avoid an activity altogether?

Here is a specific example to consider: Have you ever used mind reading to anticipate a criticism or comment, and then went so far as to create a whole dialogue for how you would respond, playing it out in your head in detail, and then the criticism or comment never even occurred? If so, write down the example. Did you put yourself through a lot of unnecessary anxiety?

Consider healthier ways of thinking about upcoming situations. Recognize when mind reading occurs and replace it with coping thoughts like I don't know for sure what anyone is thinking; it's hard to predict. In the past, I have thought that others were thinking negatively about me and there was no proof of it. I am only increasing my anxiety if I focus on critical thoughts from others that may or may not even occur.

Note that mind reading is different than being well acquainted with a person's perspectives. Sometimes you are well aware of a certain person's thoughts, and you are not mind reading, but recalling what was said in the past. For example, a friend may not want to hear talk about politics, and has said so. In this case, you can use what you know about the person's opinions to guess fairly accurately what their thoughts will be if you discuss politics, but note that when you interact with that person, it is still your decision whether you will base your behavior on your own perspective or the other person's.

Now that you know the negative impact mind reading has on your amygdala, try approaching situations without mind reading. Have an openness and curiosity toward what will happen, and work to build your comfort with living with the very real uncertainty of what others actually think unless they share it honestly. Over time, after practice in approaching a situation with curiosity rather than inaccurate assumptions and guesses, you will learn that attempts at mind reading are not always helpful and that it's better to wait to let others tell you what they are thinking. You will be surprised how often no one cares as much as you expect, or that they are neutral or positive toward your actions or statements.

44 The Trap of What-If

You are an intelligent and creative person and have the ability to imagine things that may never happen. But do you use this ability in a helpful way? You may ask yourself what-if questions and allow yourself to think the worst about what could happen. Perhaps you learned this from someone who thought it helpful to ask you questions, like "What if your car breaks down?" "What if she gets angry at you?" "What if you lose your job?" Or maybe you feel that your brain has always been focused on generating ideas about things that might occur, but which are unlikely. *What if that sound I heard was me driving over an animal? What if I did something terrible that I don't remember when I was drinking last night? What if my grandchild gets injured while I am babysitting her?* You can focus on answering these questions, coming up with more and more situations that could occur, and considering worse outcomes as your imagination suggests answers.

If you have become trapped in constantly thinking about what-ifs, your cortex is creating a great deal of misery for you. Each question you raise prompts you to imagine negative things that could occur, and new and frightening ideas roll out of your cortex regularly, like frightening nightmares that torture you. The torture comes from the fact that these ideas activate your amygdala to produce a great deal of distressing anxiety.

Regardless of how you have gotten into this practice, you must recognize how torturous it is. Perhaps it not only affects you, but you recruit others into the practice by asking them what-if questions, inviting them to help you come up with ideas of what could go wrong. Perhaps you seek reassurance from others but find that you can't always find comfort in what is said.

These questions are driven by a need to know "for sure" what could happen in the future, as if that is possible. The discomfort that many of us feel with uncertainty keeps us asking what-if questions in the hopes that we or someone else can provide answers that will provide relief. We keep turning the question over and over in our minds and trying to find answers that get us somewhere, like a person constantly turning and twisting a Rubik's cube and trying to make the colors line up correctly. But you can't find solid answers about the future no matter how you focus on these questions. You are being lured into focusing on a task that won't take you into any solid reality; it will only take you on a ride through a chamber of horrors.

Opting Out of the What-If Cycle

Take a few moments to consider situations in which you often find yourself focused on what-ifs and write those what-if questions down in your journal so that you can be watchful for them. After you have identified some to watch for, try the following steps.

1. Recognize what-if questions when they occur. If you want to get off this torturous ride, you need to detect when what-if questions come into your thoughts and understand that they are going to try to hook you into taking the bait. Spending your time answering these questions is an anxiety trap that will not help you in the way worry is intended to be helpful (see chapter 36). You are being invited to imagine problems, not solutions. Further, imagining what could happen does not make it happen, nor does it keep it from happening. It is a torturous exercise, like taking your amygdala to a horror movie.

2. Change the game. Change what-if questions to statements. *What if I don't have money for rent?* gets turned into *I may or may not have money for rent. What if Janice is angry at me?* becomes *Janice may or may not be angry at me.* You are acknowledging the uncertainty of the situation straight on. You are stating that you do not know for certain and that is that. The idea is to change the game from finding answers to being comfortable with the fact that life is uncertain, something you live with every day. Right now, for example, you don't know exactly where all your loved ones are or know with complete certainty they are all safe. If you start asking, *What if something has happened to ____*, you're inviting distress without evidence one way or the

other. Generating worrisome thoughts simply activates your amygdala and brings you anxiety.

3. Move on with your day, allowing the uncertainty of that situation to exist, and practicing living with it rather than generating frightening scenarios that only replace the uncertainty with anxiety.

When you realize that following the what-if pathway does not reduce uncertainty, but simply increases anxiety about what could potentially occur, you stop wanting to spend time spinning out various possibilities in answer to what-if questions. The future could turn out in various ways; uncertainty is part of life. Dwelling on potential negative outcomes does not make the uncertainty go away and it increases your suffering. Sometimes you just need to live with the uncertainty, especially when the situation is not something you can affect.

45 Fighting Pessimism

When your life has been dominated by fears and anxiety, for whatever reason, this shapes your thinking processes. You become watchful for signs of potential distressing situations. You get discouraged about your future and begin to believe that it is safest to expect the worst. Sometimes you can get caught up in repeatedly analyzing situations, as if to prepare for what disasters could befall you, an experience called *anxious apprehension* (Engels et al. 2007). When you think in these ways, the tendency to do so becomes stronger and stronger. Remember that the circuits in our cortex operate on the principle of "survival of the busiest" (Schwartz and Begley 2003, 17), and whatever circuitry you use repetitively is likely to be strengthened and more easily activated in the future. Negative anticipation dominates your cortex, and you are now caught in a cycle of pessimism.

Pessimism has extremely negative effects on your life. Focusing on negative thoughts and anticipating negative outcomes activates your amygdala, causing you to live in a state of anxiety and dread. Pessimism also discourages you from trying to make changes in your life. In fact, pessimism can become so strong that it discourages you from doing anything at all. Every potential action you consider can be shut down by anticipating how it can go wrong. The paralyzing effect of pessimism, combined with the negative emotions that it generates, makes a pessimistic state very debilitating.

Fighting pessimism can be difficult when it has become routine. When you try to think in more neutral or positive ways, you recognize that your cortex is tuned to focus on the negative. You find it easy to come up with potential problems and reasons why things will not work out. To focus on the positive seems foreign at first. Negative thoughts are so readily at hand, due to your previous repetitive focus on them, making them some of the strongest thoughts you have. Here are some of the most discouraging and limiting negative thoughts a person can have:

If something can go wrong, it will.

Expect the worst, and you won't be disappointed.

I'm often convinced that my troubles will never end.

Most people will let you down, so it's best not to expect much.

When these types of discouraging thoughts have taken root, you need to *fight* your pessimism.

Pessimism is more associated with the right hemisphere, but your left hemisphere is more able to create an optimistic perspective (Hecht 2013). Deliberately practicing taking a more positive view of a situation has been shown to activate the left hemisphere (McRae et al. 2012), which is evidence that a pessimistic attitude can be modified. People who are optimistic tend to be happier, handle difficulties better, and even have better health (Peters et al. 2010). Also, optimistic thinking is likely to increase your motivation to do various things despite setbacks (Sharot 2011) which leads to more resilience in life. In addition, focusing on the negative simply is not an emotionally rewarding life to live. Focusing on the Pessimism Channel is a miserable path to nowhere, and trying another approach can make your life better on so many levels.

Getting Off the Pessimism Channel

Getting off the Pessimism Channel does not begin with a focus on the pessimistic thoughts listed above, and the attempt to argue against these thoughts. That is still staying on the Pessimism Channel. To successfully reduce pessimism, you need to get off the channel completely by focusing on other ideas.

The best thing about this solution to pessimism is that it means you can deliberately focus on anything that you enjoy. You shift your focus to something that interests or entertains you and make sure you include it in your day. Whether it is chatting with a friend, playing a game on your phone, baking or tinkering, include something enjoyable in your day.

Of course, you don't need to focus on only enjoyable activities. Any change in focus away from pessimistic thoughts, even a focus on work or household tasks that you need to get done, can be an improvement. Just remember to include a focus on what you enjoy. This is encouraging, relief-producing, and more likely to help you shift your thinking.

Increasing gratitude is a particularly beneficial focus. Deliberately asking yourself each day what you are thankful for and what or whom you appreciate can dramatically improve your focus on the positive. Focus for a few moments on these phrases each day. Get out your journal and try to write new sentences each day to increase your focus on the positive.

Something that makes me smile is…

I'm thankful that…

I really appreciate when [name] [does this}…

I'm happy that I can…

Something I appreciated today that I [saw, heard, tasted, smelled, etc.] was…

It is helpful to remember that whatever you focus on becomes your reality. If you focus on negative aspects of life the majority of the time, your daily life will seem dark, dreary, and sad. When you make an effort to focus on anything that you enjoy, what makes you feel productive, or what makes you smile, your daily life becomes more engaging and enjoyable. Your focus makes a great deal of difference.

46 Stop Comparing Yourself to Others

Perhaps it feels natural to compare yourself to other people. You want to understand whether your behavior is normal, or to see if your performance at some task is adequate. When you see that someone else has trouble remembering names, it may help you feel better that you aren't the only one who experiences this difficulty. When you are discussing what to make for a potluck supper, and someone suggests lasagna because yours is the best of all the people at church, you feel some well-deserved pride. But comparing yourself to others also has many negative consequences.

Especially when you deal with anxiety difficulties, it does not make sense to compare yourself to others. When you know that you are a worrier, or that you tend to battle social anxiety, you experience challenges that are invisible to others, but which add complications to many aspects of your life. Understanding the processes in the brain that create anxiety and the inherited and learned aspects of each person's experience of anxiety, should show you that we are all unique. Each of us approaches a situation with our own challenges and strengths, and anxiety is a part of everyone's life in one way or another. Comparing oneself to others is not helpful.

For some people, anxiety makes any social situation a challenge. Being in a wedding or needing to give a presentation is an overwhelming experience. Other people can jump up and sing impromptu in front of a group without a worry about all the eyes on them. We remind our clients that even though the majority of people drive to work through busy traffic without much distress, some of our clients deserve a medal just for pushing through their anxiety to get to the office. Other people don't need to know the specific challenges you face, either because of anxiety difficulties that run in your family, or because of traumatic experiences that have made certain experiences very anxiety-provoking. But *you* need to know that sometimes you really deserve a standing ovation for something that may seem easy to other people.

We have very little control over the factors that have contributed to our challenges with anxiety, but we do have the option of taking advantage of all the ways to cope with anxiety. One good thing about having an anxiety disorder is that we do understand what happens in the brain to produce anxiety difficulties and how to rewire the brain. We don't have this level of understanding of what causes depression, bipolar disorder, autism, or dementia. At least we can discuss the parts of the brain we are trying to modify.

Comparing yourself to others is not fair to yourself or to others. In the poem "Desiderata," Max Ehrmann (1927) advises, "If you compare yourself with others, you may become vain or bitter, for always there will be greater and less persons than yourself." He also encourages you to "enjoy your achievements" and notes only you know what they have cost you. If you can catch yourself when you are comparing yourself to others, and notice the effect it has on you, we think you will agree that is not helpful to do so. Especially in the age of social media, it often appears as if others are more successful or having more fun than we are.

Watching the Effects of Comparisons

For the next day or so, be aware of when you compare yourself to others, especially when it comes to your performance of certain activities. Watch for when you see someone accomplishing something and compare yourself in a negative way. Do you criticize your own abilities or performance, assuming that you and the other person are essentially the same and should perform in the same way?

Also watch for when you see someone performing some behavior or handling a situation in a way that you think is inferior to the way you would behave. Do you use this observation to build your own self-confidence? Is it necessary for you to feel superior to others in order to have confidence in yourself? Jot down your observations in your journal before moving on to the next chapter.

Try to refocus yourself on comparing yourself with yourself. In your journal, compare how you are performing today in comparison with other times you have performed this activity. Look to see if you can see improvements in performance or the ability to do the task with less anxiety. Also, in your journal, consider your goals for yourself with what you are accomplishing. Your goals are likely to be different from others' goals, and you want to be evaluating yourself based on your own goals, rather than someone else's goals.

In your life, you will always be able to find people who do something better than you do but that is true of all of us. Similarly, we each have strengths or tendencies that others may admire and envy. Comparing yourself to others doesn't make much sense when you consider the uniqueness of each individual and how different our circumstances and experiences are.

47 Don't Should on Yourself

As we identify amygdala-activating thoughts, we can't forget to mention *should statements*. These are thoughts that are focused on the way that things "should" be. For example, *I should keep my house cleaner* or *Store clerks should not be impatient with customers*. While these statements may seem harmless, it is the "should" that can cause problems. When we say "should" it is as if we are making a rule or requirement out of a situation that is more of a preference or desire. This simple word can cause us to put increased pressure on ourselves, or to increase guilt or shame (Ellis 1987). It also can lead to increased frustration with situations and people that are beyond our control. This is why cognitive therapist Albert Ellis coined the phrase "Don't Should on Yourself."

As children, we often had "shoulds" imposed on us that may not be reasonable. Being told "You should always do your best" may seem innocent enough, but what person can really *always* do their best at every task? If you try to get up, get dressed, shower, make your bed, have breakfast, and drive to work while imposing the expectation to do every task your best, it is likely you will be exhausted by the pressure before noon. Doing our best is more appropriately tempered by other factors, including time available, our own personal needs, our personal goals, as well as a consideration of others involved.

When you think or expect that things should, must, or ought to be a certain way, you may be creating more distress for yourself. Shoulds create a distortion in your thinking because things often don't turn out as expected or hoped for, and if you focus on the way things *should* be, it becomes more likely you will be anxious, angry, guilty, or disappointed. Some expectations for ourselves and others are not realistic, and we are setting ourselves up for distress. Shoulds can be making up a rule that is not necessary. For example, if you think, *I should always do my best*, and you have a day that you were unable to, you feel defeated or guilty. Similarly, if you think, *People should not lose their temper*, you are more likely to feel distress when someone does lose their temper, as if a law had been broken. Try to identify thoughts you have that contain the words "should," "ought," or "must." Do they activate your amygdala to respond as if some kind of threat is occurring, when the impact in your life could be more minimal if you changed your expectation?

Searching for Your Own "Shoulds" and Replacing Them

Consider these questions and try to identify the shoulds in your life. Write down some examples in your journal so that you can work on replacing them with more realistic thoughts.

> Do you place shoulds on yourself, almost like a strict rule that you must not break? Do you tell

yourself you should always be early or on time to all meetings and appointments; or you should never forget a name or a birthday? The average person can't always live up to these shoulds. You are not in danger if you don't always live up to your intentions. Others are often more accepting than you are. Changing the thought to *I'll strive to be on time* or *I want to remember my friends' birthdays* can express your intention without creating the belief you are a failure or in danger if you don't live up to every good intention that you have.

Do you place shoulds on others that they are unlikely to live up to? When you believe that someone should not lose their temper, or should automatically recognize what is important to you, you may be setting an unrealistic expectation. It is alright to think, *I don't like it when people lose their temper,* or *I would appreciate it if the kids wouldn't leave their toys on the floor,* but we should not act as though some violation has occurred.

Do you place shoulds on situations that make it more difficult for you to tolerate them? When you think *I should not have gotten this illness* or *People who break traffic laws should be punished,* you are increasing your frustration or irritation. When you label it as a "should," making an annoying situation into a violation, it becomes a bigger problem than it needs to be. The way that we think about situations will influence the degree to which the amygdala reacts. To say, *I wish I had not gotten this illness* or *People who break laws annoy me,* is less activating.

Often, we learn to impose shoulds on ourselves, others, and the world without realizing that they can make our lives more frustrating and anxiety-producing than necessary. While having preferences is natural, turning them into commandments, like *You should always try your best* or *People should not lose their tempers* is not only unrealistic, but is amygdala-activating.

48 The Perils of Perfectionism

As I (Maha) sit to write, I am plagued by thoughts, like *Is this good enough? Did I write it perfectly? Can I do better, or can I do more?* You may have struggled with similar thoughts, nagging at you that you are not acting perfectly or accusing you that your work is not good enough. People with anxiety sometimes struggle with perfectionism. Perfection is defined as the need to be or appear perfect, even though it is illogical to believe that any person can achieve perfection. Perfectionists set extremely high standards and expectations for themselves and others and are often very critical of mistakes or anything that falls short of perfect. If you feel the need to be perfect in one or more areas of your life, including work, relationships, hobbies, and general pursuit of goals, you may not realize that you are increasing your daily anxiety. In fact, oftentimes perfectionism is mistaken as a positive trait. Although it seems like perfection helps you achieve, do things well, and feel special, it often comes at a huge cost. Ask yourself, *What is striving for these high standards costing me? Is the time that I have spent worth it? Is the pressure and anxiety necessary? Does it sometimes keep me from doing what I want out of fear it won't be good enough?*

Perfectionism is often associated with anxiety, high stress levels, procrastination, excessive time devoted to tasks, and a general sense of dissatisfaction. When you don't meet the high standards that you set, as no one can achieve perfection, you are set up to try more and more to improve performance such that you can't get off the hamster wheel. Judging yourself against perfection, you are constantly forced to contend with a general sense of defeat and hopelessness. You are creating a sense of danger that activates the amygdala, making yourself feel a very real threat.

Perfectionism is maintained by thinking patterns that include all-or-nothing thinking (*If I can't do it perfectly, then I won't do it at all*), catastrophic thinking (*If I don't do it perfectly, I will get an F*), and should statements (*I should be more neat and organized*). These rigid thinking patterns affect behavior. Behaviors that are motivated by perfectionism include over-doing it to meet unrelenting standards, complete avoidance because you cannot meet your high standards, and spending too much time on tasks, especially when checking or rechecking.

To reduce perfectionism and ultimately anxiety, start by being aware of your perfectionistic patterns and what areas of your life they are affecting, and recognize how they cost you. Second, strategies that help you resist both thinking patterns and behaviors will be most effective. Below are some strategies that can help.

Specific Ways to Combat Perfectionism

Take some time to read the following and write in your journal perfectionistic thoughts and, most importantly, possible replacement thoughts you could use to take the pressure off yourself.

1. Recognize and challenge rigid thoughts related to perfectionism. For example, instead of *I am a failure if I don't achieve the highest accomplishment*, adapt this thought to *I did a good job, and don't need to be so demanding of myself. I should do better on this* can be replaced with *This specific task does not require perfection; just get it done*. Thought challenging helps you reduce that critical voice always telling you it's not good enough.

2. Practice self-compassion. Talk to yourself in a way that is kind and compassionate rather than critical. Write in your journal examples of statements you think will help, like *I am only human, and mistakes are normal.*

3. Practice acting imperfectly. Identify an area in your life where you hold yourself to high, unrelenting standards. By purposely allowing yourself to do less, you will be facing the fear of not meeting strict standards and see that life goes on. Strive for "good enough" performance. Deliberately spend less time on tasks, and

check or recheck less, and you will realize that no catastrophe occurs. Deliberately send an email with a spelling or grammatical error in it, rather than correcting it, and see what happens. Have someone visit without trying to make everything spotless. The idea is that you are practicing "good enough" and seeing that the outcome is still ok. With enough practice, your amygdala learns that "good enough" is not a threat.

4. 70% rule. Perfectionism keeps you always striving for 100%. Choose certain tasks and lower your standard to 70%; this helps promote a "good is good enough" mindset and behavior in some areas. Although you may feel anxious at first, your amygdala will get used to it. After trying the "good enough" or 70% approaches, write in your journal if your life is different in certain areas after taking the pressure off.

Recognize when you have set high, unrelenting standards that lead you to either procrastinate or spend too much time. Notice how much unnecessary anxiety and stress these standards create in your life. Notice how other people do tasks and recognize that you do not need to be better than others. You can take control and reduce anxiety by embracing being human and focusing on "good enough." Save those high standards for a very limited number of tasks.

49 Abolishing Doubt Is Impossible

Many of us seek answers to our problems by thinking about them. We believe that we can figure out solutions using our logic and planning abilities. We approach our concerns like a math problem for which a single, correct answer can be found with enough calculation. This is not life. For many situations, no simple solution exists. In addition, we can't even be certain if the way we see the situation is correct. Especially when our concerns involve predicting human behavior, knowing the future, or having control of a situation, it is very unlikely that we will be able to find a single correct answer or a way to predict the outcome with certainty.

The answer to many distressing or unpredictable situations is not to try to find the answer at all. So often, finding the correct, certain answer to a question is impossible. The best approach is to increase your comfort with the uncertainty and doubts that are a normal part of life. Sometimes we need to find peace in living with the question, rather than continually seeking the answer. Considering some specific examples of doubts can help to illustrate this. We often work with clients who have difficulty resolving doubts they have about themselves, ranging from whether they ran over an animal on the way to work to if they have thoughts that are unacceptable to God.

Uncertainties plague many of our clients. *Is this egg still safe to eat? Will I be able to write a dissertation that will pass the committee? Was I ever abusive to my children? Will my husband keep working on our marriage? Should I be wearing a mask if I have a cough? Is it safe for me to drive on the highway today?* We need to admit that our lives are filled with situations in which it is often impossible to resolve doubts. We often remind our clients, when they ask us to provide them certain answers, that we cannot do so.

When clients ask us to reassure them that they will get the promotion, be able to get pregnant, or be able to attend a wedding without a panic attack, we need to be honest that we cannot provide them with certainty. But that should not stop them from living their lives as if the promotion, pregnancy, or wedding attendance is possible. We sometimes say, "I can't guarantee that you will get back home safely from my office, but that doesn't mean that you shouldn't attempt the drive." This statement is not intended to frighten them; it is intended to illustrate that we live with uncertainty all the time, and that it is better to accept it as a normal part of life.

In fact, when we realize how often we ignore uncertainties that are part of our lives, and live our lives without close attention to them, we realize that we can handle having doubts, uncertainty, and unanswered questions. One of the worst things that we can do is to ruminate on the possibilities and become obsessed with the need for an answer. Seeking reassurance from others, googling the topic, and continually dwelling on the topic in your mind is not the answer. These approaches keep the situation in your cortex, and your amygdala is constantly activated as if a threat is present. If you can accept that you have a question, leave it as a question, and live your life, you will have a calmer amygdala. Focus on what you need and want to do each day, and you will experience much less anxiety.

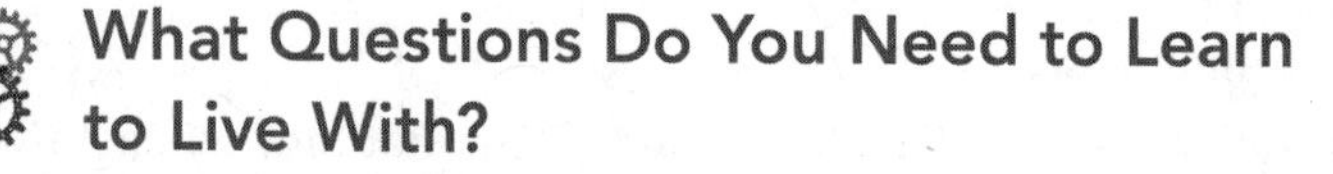

What Questions Do You Need to Learn to Live With?

Are you troubled by specific questions about your future, about others' behavior, or about risks? Remember that focusing on these questions and trying to find an answer in various ways (seeking reassurance, googling, ruminating, and incessant planning for problems) will trap you on the Anxiety Channel. Instead, identify the question, and accept the doubt and uncertainty.

Here is the kind of dialogue you should have with yourself:

I want to know [insert question]...

But I can't get an answer that I can be certain about. I need to accept that.

I can live with this question.

The best way to live with the question is to live my daily life, according to my own values and needs and interests.

Finding relief does not require having answers to all your questions. The way to a happy life is to focus your thoughts on living your life. Don't wait for answers or spend time trying to find them when certainty is not possible. Focus on what you need to do each day, and make sure that you routinely include activities that you enjoy.

50 Don't Focus on Content

Perhaps you are beginning to see an important pattern that occurs in your brain that repeatedly creates problematic anxiety. You begin to think about a situation that may create some kind of threat or negative outcome, and you ponder it. Turning the situation over and over in your mind comes very naturally. As you consider the situation, maybe you are thinking of ways to cope with it and maybe not. But this thinking process in your cortex definitely impacts your amygdala. You can feel the tell-tale signs of amygdala activation. You feel distress and discomfort in various parts of your body as the defense response kicks in. A feeling of dread develops which makes you focus your cortex-based thinking even more on the situation. It's easy to get caught in a spiral of anxiety and even panic.

Now that you recognize this pattern, it is important that you recognize that *the specifics of your thoughts don't matter.* Whether you are thinking about a possible problem in a relationship or financial difficulties, the underlying process is the same. Whenever you focus your cortex on a situation that could potentially pose a threat, your amygdala reacts, and you suffer the pangs of anxiety.

The content of the thoughts you are focused on *does not matter.* The same discomfort, distress, and anguish follow in a disturbingly predictable pattern. Once you recognize this, you can learn that *letting go of the content* can be very helpful in taking back control over your life. It doesn't matter if the content is about you potentially having cancer or the concern that you and your sister may never speak again. Whatever the content, if you can keep yourself from believing that this particular concern is something it is necessary to focus on, you can free yourself from a great deal of anxiety.

Reid Wilson (2016, 26) very wisely wrote, "As long as Anxiety can keep your attention on your content, it wins, and you will continue to lose out. Elevate the game! Get above your content, up to the real challenge you must face." Once you know that changing the focus of your thoughts is the way to stop your brain from producing anxiety, you recognize that *it doesn't matter what the specific thoughts are*—you can find freedom from anxiety by interrupting this frustrating pattern.

Can you recognize that, any day of the week, your cortex can cycle through various worries and concerns, stopping when it finds something that activates the amygdala, staying on that content, stirring the amygdala up into a frenzy, and putting your body into a state of anxiety? You can completely change the pattern if you focus your efforts on not being caught up in the content, but instead saying to yourself, *Yes, yes, you have found something else to poke the amygdala with, but I'm not going to participate. I have other things to focus on that are more important to my life*. Recognize and believe that this content, whatever it is, is *not* the real issue. The real issue is a battle over what you are going to focus on in your cortex. Are you going to dwell on and strengthen the thoughts that activate the amygdala, or shift your focus to what is important in your life at the moment?

Don't Focus on the Content

When your thoughts turn to a troubling situation, it is so easy to begin deepening your focus on that situation and get tricked into considering it in detail. But does the situation deserve that kind of attention? Sometimes this is something you have repeatedly focused on, and sometimes it is a new concern, but it is not always necessary to focus on the situation. Especially when you know you have a tendency to worry, obsess, or ruminate, you need to learn to shift your focus away from the specific content and not allow yourself to be caught up in it.

Thoughts that are signs of being caught in the content:

- *What are the possible ways this could turn out badly?*
- *How am I going to fix this situation?*
- *I don't think I can deal with this situation.*
- *This needs to be discussed with someone...*

Get out your journal and write down three to five situations in your life in which you see yourself caught in thoughts that keep you focused on the content. Now look at the next list to see if you can use these helpful thoughts to resist thoughts that activate the amygdala.

Thoughts that help you move on with your day and not get pulled into creating anxiety:

- *I don't need to focus on this; I have better things to focus on.*
- *These are only thoughts, not necessarily what will happen.*
- *I've gone over this before and can't waste more time.*
- *No amount of thinking will change this situation.*
- *I'll be able to deal with this later (if it even occurs).*
- *What do I need or want to focus on right now?*

Every day of your life you face a choice: Today, are you going to use your cortex to create anxiety, or are you going to use it to focus on what provides meaning, purpose, and pleasure in your life? Freeing yourself from focusing on the content means that you are the one who directs your life.

References

Altman, D. 2014. *The Mindfulness Toolbox.* Eau Claire, WI: PESI.

Anderson, E., and G. Shivakumar. 2013. "Effects of Exercise and Physical Activity on Anxiety." *Frontiers in Psychiatry* 4(27): 1–4.

Babson, K. A., M. T. Feldner, C. D. Trainor, and R. C. Smith. 2009. "An Experimental Investigation of the Effects of Acute Sleep Deprivation on Panic-Relevant Biological Challenge Responding." *Behavior Therapy* 40(3): 239–250.

Balaghi, D., K. Hierl, and E. Snyder. 2022. "Self-Monitoring for Students with Obsessive-Compulsive Disorder and Autism Spectrum Disorder." *Intervention in School and Clinic* 58(1): 51–58.

Bourne, E. J., A. Brownstein, and L. Garano. 2004. *Natural Relief for Anxiety: Complementary Strategies for Easing Fear, Panic, and Worry.* Oakland, CA: New Harbinger.

Burton, L. R. 2020. "The Neuroscience and Positive Impact of Gratitude in the Workplace." *The Journal of Medical Practice Management* 35(4): 215–218.

Chang, A., D. Aeschbach, J. F. Duffy, and C. A. Czeisler. 2015. "Evening Use of Light-Emitting eReaders Negatively Affects Sleep, Circadian Timing, and Next-Morning Alertness." *Proceedings of the National Academy of Sciences of the United States of America* 112(4): 1232–1237.

Chen, Y., C. Chen, R. M. Martínez, J. L. Etnier, and Y. Cheng. 2019. "Habitual Physical Activity Medicates the Acute Exercise-Induced Modulation of Anxiety-Related Amygdala Functional Connectivity." *Scientific Reports* 9(1): 19787.

Christianson, J. P., and B. N. Greenwood. 2014. "Stress-Protective Neural Circuits: Not All Roads Lead Through the Prefrontal Cortex." *Stress* 17(1): 1–12.

DeBoer, L. B., M. B. Powers, A. C. Utschig, M. W. Otto, and J. A. J. Smits. 2012. "Exploring Exercise as an Avenue for the Treatment of Anxiety Disorders." *Expert Review of Neurotherapeutics* 12(8): 1011–1022.

Ehrmann, M. 1948. *The Poems of Max Ehrmann.* Edited by Bertha K. Ehrmann. Boston: Bruce Humphries.

Ellis, A. 1987. "A Sadly Neglected Cognitive Element in Depression." *Cognitive Therapy and Research* 11(1): 121–145.

Engels, A. S., W. Heller, A. Mohanty, J. D. Herrington, M.T. Banich, A. G. Webb, et al. 2007. "Specificity of Regional Brain Activity in Anxiety Types During Emotion Processing." *Psychophysiology* 44(3): 352–363.

Ensari, I., T. A. Greenlee, R. W. Motl, and S. J. Petruzzello. 2015. "Meta-Analysis of Acute Exercise Effects on State Anxiety: An Update of Randomized Controlled Trials Over the Past 25 years." *Depression and Anxiety* 32(8): 624–634.

Erikson, E. H. 1963. *Childhood and Society.* New York: W. W. Norton.

Goldin, P. R., and J. J. Gross. 2010. "Effects of Mindfulness-Based Stress Reduction (MBSR) on Emotion Regulation in Social Anxiety Disorder." *Emotion* 10(1): 83–91.

Harris, R. 2008. *The Happiness Trap: How to Stop Struggling and Start Living.* Boulder, CO: Trumpeter.

Hecht, D. 2013. "The Neural Basis of Optimism and Pessimism." *Experimental Neurobiology* 22(3): 173–199.

Heisler, L. K., L. Zhou, P. Bajwa, J. Hsu, and L. H. Tecott. 2007. "Serotonin 5-HT2c Receptors Regulate Anxiety-Like Behavior." *Genes, Brain, and Behavior* 6(5): 491–496.

Lattari, E., H. Budde, F. Paes, G. A. M. Neto, J. C. Appolinario, A. E. Nardi, et al. 2018. "Effects of Aerobic Exercise on Anxiety Symptoms and Cortical Activity in Patients with Panic Disorder: A Pilot Study." *Clinical Practice and Epidemiology in Mental Health* 14: 11–25.

LeDoux, J. 2015. *Anxious: Using the Brain to Understand and Treat Fear and Anxiety.* New York: Penguin.

Linehan, M. M. 2014. *DBT Skills Training Manual.* 2nd ed. New York: Guilford Press.

Lombardi, G., M. Gerbella, M. Marchi, A. Sciutti, G. Rizzolatti, and G. Di Cesare. 2023. "Investigating Form and Content of Emotional and Non-Emotional Laughing." *Cerebral Cortex* 33(7): 1–9.

McCullough, M. E., R. A. Emmons, and J. Tsang. 2002. "The Grateful Disposition: A Conceptual and Empirical Topography." *Journal of Personality and Social Psychology* 82(1): 112–127.

McNay, E. 2015. "Recurrent Hypoglycemia Increases Anxiety and Amygdala Norepinephrine Release During Subsequent Hypoglycemia." *Frontiers in Endocrinology* 6: 175.

McRae, K., J. J. Gross, J. Weber, E. R. Robertson, P. Sokol-Hessner, R. D. Ray, et al. 2012. "The Development of Emotion Regulation: An fMRI Study of Cognitive Reappraisal in Children, Adolescents, and Young Adults." *Social Cognitive and Affective Neuroscience* 7(1): 11–22.

Menzies, V., D. E. Lyon, R. K. Elswick Jr., N. L. McCain, and D. P. Gray. 2014. "Effects of Guided Imagery on Biobehavioral Factors in Women with Fibromyalgia. *Journal of Behavioral Medicine* 37(1): 70–80.

Neff, K. D. 2023. "Self-Compassion: Theory, Method, Research, and Intervention." *Annual Review of Psychology* 74: 193–218.

Orsillo, S. M., L. Roemer, J. B. Lerner, and M. T. Tull. 2004. "Acceptance, Mindfulness, and Cognitive-Behavioral Therapy: Comparisons, Contrasts, and Application to Anxiety." In *Mindfulness and Acceptance: Expanding the Cognitive-Behavioral Tradition*, edited by S. C. Hayes, V. M. Follette, and M. M. Linehan. New York: Guilford Press.

Perreau-Linck, E., M. Beauregard, P. Gravel, V. Paquette, J. Soucy, M. Diksic, et al. 2007. "In Vivo Measurements of Brain Trapping of C-Labelled α-Methyl-L-Tryptophan During Acute Changes in Mood States." *Journal of Psychiatry Neuroscience* 32(6): 430–434.

Peters, M. L., I. K. Flink, K. Boersma, and S. J. Linton. 2010. "Manipulating Optimism: Can Imagining a Best Possible Self Be Used to Increase Positive Future Expectancies?" *The Journal of Positive Psychology* 5(3): 204–211.

Pilozzi, A., C. Carro, and X. Huang. 2021. "Roles of ß-Endorphin in Stress, Behavior, Neuroinflammation, and Brain Energy Metabolism." *International Journal of Molecular Sciences* 22(1): 338.

Prather, A. A., R. Bogdan, and A. R. Hariri. 2013. "Impact of Sleep Quality on Amygdala Reactivity, Negative Affect, and Perceived Stress." *Psychosomatic Medicine* 75(4): 350–358.

Rebar, A. L., R. Stanton, D. Geard, C. Short, M. J. Duncan, and C. Vandelanotte. 2015. "A Meta-Meta-Analysis of the Effect of Physical Activity on Depression and Anxiety in a Non-Clinical Adult Population." *Health Psychology Review* 9(3): 366–378.

Roehrs, T., and T. Roth. 2008. "Caffeine: Sleep and Daytime Sleepiness." *Sleep Medicine Reviews* 12(2): 153–162.

Rogers, P. J. 2007. "Caffeine, Mood, and Mental Performance in Everyday Life." *Nutrition Bulletin* 32: 84–89.

Sander, K., A. Brechmann, and H. Scheich. 2003. "Audition of Laughing and Crying Leads to Right Amygdala Activation in a Low-noise fMRI Setting." *Brain Research Protocols* 11(2): 81–91.

Schwartz, J. M., and S. Begley. 2003. *The Mind and the Brain: Neuroplasticity and the Power of Mental Force.* New York: Harper Perennial.

Schmitt, A., N. Upadhyay, J. A. Martin, S. R. Vega, H. K. Strüder, and H. Boecker. 2020. "Affective Modulation After High-Intensity Exercise is Associated with Prolonged Amygdala-Insular Functional Connectivity Increase." *Neural Plasticity* 2020(5): 1–10.

Sharot, T. 2011. "The Optimism Bias." *Current Biology* 21(23): R941–945.

Summer, J. 2023. "What is REM Sleep and How Much Do You Need?" *Sleep Foundation*, May 9. https://www.sleepfoundation.org/stages-of-sleep/rem-sleep.

Taylor, V. A., J. Grant, V. Daneault, G. Scavone, E. Breton, S. Roffe-Vidal, et al. 2011. "Impact of Mindfulness on the Neural Responses to Emotional Pictures in Experienced and Beginner Meditators." *Neuroimage* 57(4): 1524–1533.

van der Helm, E., J. Yao, S. Dutt, V. Rao, J. M. Salentin, and M. P. Walker. 2011. "REM Sleep De-Potentiates Amygdala Activity to Previous Emotional Experiences." *Current Biology* 21(23): 2029–2032.

Watson, E. J., A. M. Coates, M. Kohler, and S. Banks. 2016. "Caffeine Consumptions and Sleep Quality in Australian Adults." *Nutrients* 8(8): 479.

Woodbury-Fariña, M., and M. M. R. Schwabe. 2015. "Laughter Yoga: Benefits of Mixing Laughter and Yoga." *Journal of Yoga & Physical Therapy* 5(4): 209.

Wilson, R. 2016. *Stopping the Noise in Your Head: The New Way to Overcome Anxiety and Worry.* Deerfield Beach, FL: Health Communications.

Yim, J. 2016. "Therapeutic Benefits of Laughter in Mental Health: A Theoretical Review." *Tohoku Journal of Experimental Medicine* 239(3): 243–249.

Yoo, S., N. Gujar, P. Hu, F. A. Jolesz, and M. P. Walker. 2007. "The Human Emotional Brain Without Sleep—A Prefrontal Amygdala Disconnect." *Current Biology* 17(20): R877–878.

Zelano, C., H. Jiang, G. Zhou, N. Arora, S. Schuele, J. Rosenow, et al. 2016. "Nasal Respiration Entrains Human Limbic Oscillations and Modulates Cognitive Function." *The Journal of Neuroscience* 36(49): 12448–12467.

Catherine M. Pittman, PhD, is a licensed clinical psychologist specializing in the treatment of anxiety disorders and brain injuries. She is professor of psychology at Saint Mary's College in Notre Dame, IN, where she has taught for more than thirty years. Pittman is author of *Taming Your Amygdala*, and coauthor of *Rewire Your Anxious Brain*.

Maha Zayed Hoffman, PhD, is a licensed clinical psychologist specializing in the treatment of anxiety and obsessive-compulsive disorder (OCD). She is owner of The OCD & Anxiety Center, with offices in Oak Brook, IL; Orland Park, IL; and Marietta, GA.